Notes on Counting: An Introduction to Enumerative Combinatorics

Enumerative combinatorics, in its algebraic and analytic forms, is vital to many areas of mathematics, from model theory to statistical mechanics. This book, which stems from many years' experience of teaching, invites students into the subject and prepares them for more advanced texts. It is suitable as a class text or for individual study.

The author provides proofs for many of the theorems to show the range of techniques available and uses examples to link enumerative combinatorics to other areas of study. The main section of the book introduces the key tools of the subject (generating functions and recurrence relations), which are then used to study the most important combinatorial objects, namely subsets, partitions, and permutations of a set. Later chapters deal with more specialised topics, including permanents, SDRs, group actions and the Redfield–Pólya theory of cycle indices, Möbius inversion, the Tutte polynomial, and species.

G000068239

Australian Mathematical Society Lecture Series: 26

Notes on Counting: An Introduction to Enumerative Combinatorics

PETER J. CAMERON

University of St Andrews, Scotland

CAMBRIDGE
UNIVERSITY PRESS

CAMBRIDGE
UNIVERSITY PRESS

University Printing House, Cambridge CB2 8BS, United Kingdom

One Liberty Plaza, 20th Floor, New York, NY 10006, USA

477 Williamstown Road, Port Melbourne, VIC 3207, Australia

4843/24, 2nd Floor, Ansari Road, Daryaganj, Delhi – 110002, India

79 Anson Road, #06-04/06, Singapore 079906

Cambridge University Press is part of the University of Cambridge.

It furthers the University's mission by disseminating knowledge in the pursuit of education, learning and research at the highest international levels of excellence.

www.cambridge.org

Information on this title: www.cambridge.org/9781108417365

DOI: 10.1017/9781108277457

First published 2017

Printed in the United Kingdom by Clays, St Ives plc

A catalogue record for this publication is available from the British Library.

ISBN 978-1-108-41736-5 Hardback

ISBN 978-1-108-40495-2 Paperback

Contents

Preface

Combinatorics is the science of pattern and arrangement. A typical problem in combinatorics asks whether it is possible to arrange a collection of objects according to certain rules. If the arrangement is possible, the next question is a counting question: how many different arrangements are there? This is the topic of the present book.

Often a small change in the detail of a problem turns an easy question into one which appears impossibly difficult. For example, consider the following three questions.

- In how many ways can the numbers $1, \ldots, n$ be placed in the cells of an $n \times n$ grid, with no restriction on how many times each is used? Since each of the n^2 cells can have its entry chosen independently from a set of n possibilities, the answer is n^{n^2}.

- In how many ways can the arrangement be made if each number must occur once in each row? Once we notice that each row must be a permutation of the numbers $1, \ldots, n$, and that the permutations can be chosen independently, we see that the answer is $(n!)^n$ (as there are $n!$ permutations of the numbers $1, \ldots, n$).

- In how many ways can the arrangement be made if each number must occur once in each row and once in each column? For this problem, there is no formula for the answer. Such an arrangement is called a *Latin square*. The number of Latin squares with n up to 11 has been found by brute-force calculation. For larger values, we don't even have good estimates: the best known upper and lower bounds differ by a factor which is exponentially large in terms of the number of cells.

Not all problems are as hard as this. In this book you will learn how to count

ix

the number of permutations which move every symbol, strings of zeros and ones containing no occurrence of a fixed substring, invertible matrices of given size over a finite field, expressions for a given positive integer as a sum of positive integers (the answers are different depending on whether we care about the order of the summands or not), trees and graphs on a given set of vertices, and many more.

Very often I will not be content with giving you one proof of a theorem. I may return to an earlier result armed with a new technique and give a totally different proof. We learn something from having several proofs of the same result. As Michael Atiyah said in an interview in the *Newsletter of the European Mathematical Society*,

> I think it is said that Gauss had ten different proofs for the law of quadratic reciprocity. Any good theorem should have several proofs, the more the better. For two reasons: usually, different proofs have different strengths and weaknesses, and they generalise in different directions — they are not just repetitions of each other.

In particular, there are two quite different styles of proof for results in enumerative combinatorics. Consider a result which asserts that two counting functions $F(n)$ and $G(n)$ are equal. We might prove this by showing that their generating functions are equal, perhaps using analytic techniques of some kind. Alternatively, we might prove the result by finding a bijection between the sets of objects counted by $F(n)$ and $G(n)$. Often, when such an identity is proved by analytic methods, the author will ask for a 'bijective proof' of the result.

As an example, if n is even, then it is fairly straightforward to prove by analytic methods that the number of permutations of $\{1, \ldots, n\}$ with all cycles even is equal to the number with all cycles odd. But finding an explicit bijection between the two sets is not straightforward, though not too difficult.

I should stress, though, that the book is not full of big theorems. Tim Gowers, in a perceptive article on 'The two cultures of mathematics', distinguishes branches of mathematics in which theorems are all-important from those where the emphasis is on techniques; enumerative combinatorics falls on the side of techniques. (In the past this has led to some disparagement of combinatorics by other mathematicians. Many people know that Henry Whitehead said 'Combinatorics is the slums of topology'. A more honest appraisal is that the techniques of combinatorics pervade all of mathematics, even the most theorem-rich parts.)

The notes which became this book were for a course on *Enumerative and Asymptotic Combinatorics* at Queen Mary, University of London, in the spring of 2003, and subsequently as *Advanced Combinatorics* at the University of St Andrews. The reference material for the subject has been greatly expanded by the publication of Richard Stanley's two-volume work on *Enumerative Combinatorics*, as well as the book on *Analytic Combinatorics* by Flajolet and Sedgewick. (References to these and many other books can be found in the bibliography at the end.) Many of these books are encyclopaedic in nature. I hope that this book will be an introduction to the subject, which will encourage you to look further and to tackle some of the weightier tomes.

What do you need to know to read this book? It will probably help to have had some exposure to basic topics in undergraduate mathematics.

- Real and complex analysis (limits, convergence, power series, Cauchy's theorem, singularities of complex functions);
- Abstract algebra (groups and rings, group actions);
- Combinatorics.

None of this is essential; in most cases you can pick up the needed material as you go along.

The heart of the book is Chapters 2–4, in which the most important tools of the subject (generating functions and recurrence relations) are introduced and used to study the most important combinatorial objects (subsets, partitions and permutations of a set). The basic object here is a *formal power series*, a single object encapsulating an infinite sequence of numbers, on which a wide variety of manipulations can be done: formal power series are introduced in Chapter 2.

Later chapters treat more specialised topics: permanents, systems of distinct representatives, and Latin squares in Chapter 5, 'q-analogues' (familiar formulae with an extra parameter arising in a wide variety of applications) in Chapter 6, group actions and the Redfield–Pólya theory of the cycle index in Chapter 7, Möbius inversion (a wide generalisation of the Inclusion–Exclusion Principle) in Chapter 8, the Tutte polynomial (a counting tool related to Inclusion–Exclusion) in Chapter 9, species (an abstract formalism including many important counting problems) in Chapter 10, and some miscellaneous topics (mostly analytic) in Chapters 11 and 12. The final chapter includes an annotated list of books for further study.

As always in a combinatorics book, the techniques described have unexpected applications, and it is worth looking through the index. Cayley's Theorem on trees, for example, appears in Chapter 10, where several different proofs are given; Young tableaux are discussed in Chapter 4, as are various counts of inverse semigroups of partial permutations.

The final chapter includes an annotated book list and a discussion of using the On-line Encyclopedia of Integer Sequences.

The first few chapters contain various interdependences. For example, binomial coefficients and the Binomial Theorem for natural number exponents appear in Chapter 2, although they are discussed in more detail in Chapter 3. Such occurrences will be flagged, but I hope that you will have met these topics in undergraduate courses or will be prepared to take them on trust when they first appear.

I am grateful to many students who have taken this course (especially Pablo Spiga, Thomas Evans, and Wilf Wilson), to colleagues who have helped teach it (especially Thomas Müller), and to others who have provided me with examples (especially Thomas Prellberg and Dudley Stark), and to Abdullahi Umar for the material on inverse semigroups. I am also grateful to Morteza Mohammed-Noori, who used my course notes for a course of his own in Tehran, and did a very thorough proof-reading job, spotting many misprints. (Of course, I may have introduced further misprints in the rewriting!)

The book will be supported by a web page at

```
http://www-circa.mcs.st-and.ac.uk/~pjc/books/counting/
```

which will have a list of misprints, further material, links, and possibly solutions to some of the exercises.

CHAPTER 1

Introduction

This book is about counting. Of course this doesn't mean just counting a single finite set. Usually, we have a family of finite sets indexed by a natural number n, and we want to find $F(n)$, the cardinality of the nth set in the family. For example, we might want to count the subsets or permutations of a set of size n, lattice paths of length n, words of length n in the alphabet $\{0, 1\}$ with no two consecutive 1s, and so on.

1.1 What is counting?

There are several kinds of answer to this question:

- An explicit formula (which may be more or less complicated, and in particular may involve a number of summations). In general, we regard a simple formula as preferable; replacing a formula with two summations by one with only one is usually a good thing.

- A recurrence relation expressing $F(n)$ in terms of n and the values of $F(m)$ for $m < n$. This allows us to compute $F(0), F(1), \ldots$ in turn, up to any desired value.

- A closed form for a *generating function* for F. We will have much more to say about generating functions later on. Roughly speaking, a generating function represents a sequence of numbers by a power series, which in some cases converges to an analytic function in some domain in the complex plane. An explicit formula for the generating function for a sequence of numbers is regarded as almost as good as a formula for the numbers themselves.

1

If a generating function converges, it is possible to find the coefficients by analytic methods (differentiation or contour integration).

In the examples below, we use two forms of generating function for a sequence $(a_0, a_1, a_2, \ldots)$ of natural numbers: the *ordinary generating function*, given by

$$\sum_{n \geq 0} a_n x^n,$$

and the *exponential generating function*, given by

$$\sum_{n \geq 0} \frac{a_n x^n}{n!}.$$

We will study these further in the next chapter, and meet them many times during later chapters. In Chapter 10, we will see a sort of explanation of why some problems need one kind of generating function and some need the other.

- An asymptotic estimate for $F(n)$ is a function $G(n)$, typically expressed in terms of the standard functions of analysis, such that $F(n) - G(n)$ is of smaller order of magnitude than $G(n)$. (If $G(n)$ does not vanish, we can write this as $F(n)/G(n) \to 1$ as $n \to \infty$.) We write $F(n) \sim G(n)$ if this holds. This might be accompanied by an asymptotic estimate for $F(n) - G(n)$, and so on; we obtain an *asymptotic series* for F. (The basics of asymptotic analysis are described further in the next section of this chapter.)

- Related to counting combinatorial objects is the question of generating them. The first thing we might ask for is a system of sequential generation, where we can produce an ordered list of the objects. Again there are two possibilities.

  If the number of objects is $F(n)$, then we can in principle arrange the objects in a list, numbered $0, 1, \ldots, F(n) - 1$; we might ask for a construction which, given i with $0 \leq i \leq F(n) - 1$, produces the ith object on the list directly, without having to store the entire list and count through from the start.

  Alternatively, we may simply require a method of moving from each object to the next.

- We could also ask for a method for random generation of an object. If we have a technique for generating the ith object directly, we simply choose a random number in the range $\{0, \ldots, F(n) - 1\}$ and generate the corresponding object. If not, we have to rely on other methods such as Markov chains.

Here are a few examples. These will be considered in more detail in Chapter 3; it is not necessary to read what follows here in detail, but you are advised to skim through it.

Example: subsets   The number of subsets of $\{1, \ldots, n\}$ is 2^n. For each subset is specified by saying, for each number $i \in \{1, \ldots, n\}$, whether i is in the subset or not; thus n binary choices are required to specify the subset.

Not only is this a simple formula to write down; it is easy to compute as well. It can clearly be done by starting with 1 and doubling n times (that is, n integer additions). Alternatively, it can be computed with at most $2\log_2 n$ integer multiplications. (In other words, we can choose to have fewer but more complicated operations.)

To see this, write n in base 2: $n = 2^{a_1} + 2^{a_2} + \cdots + 2^{a_r}$, where $a_1 > \cdots > a_r$. Now we can compute 2^{2^i} for $1 \le i \le a_1$ by a_1 successive squarings (noting that $2^{2^{i+1}} = \left(2^{2^i}\right)^2$); then $2^n = (2^{2^{a_1}}) \cdots (2^{2^{a_r}})$ requires $r - 1$ further multiplications.

There is a simple recurrence relation for $F(n) = 2^n$, namely

$$F(0) = 1, \qquad F(n) = 2F(n-1) \text{ for } n \ge 1.$$

This expresses the calculation of 2^n by n doublings. Another recurrence relation, expressing the more efficient technique just outlined for computing 2^n, is given by

$$F(0) = 1, \qquad F(n) = \begin{cases} 2F(n-1) & \text{if } n \text{ is odd,} \\ F(n/2)^2 & \text{if } n \text{ is even.} \end{cases}$$

The ordinary generating function of the sequence (2^n) is

$$\sum_{n \ge 0} 2^n x^n = \frac{1}{1 - 2x},$$

while the exponential generating function is

$$\sum_{n \ge 0} \frac{2^n x^n}{n!} = \exp(2x).$$

(I will use $\exp(x)$ instead of e^x in these notes, except in some places involving calculus.)

No asymptotic estimate is needed, since we have a simple exact formula. Indeed, it is clear that 2^n is a number with $\lceil n \log_{10} 2 \rceil$ decimal digits.

Choosing a random subset, or generating all subsets in order, are easily achieved by the following method. For each $i \in \{0, \ldots, 2^n - 1\}$, write i in base 2, producing a string of length n of zeros and ones. Now j belongs to the ith subset if and only if the jth symbol in the string is 1.

A procedure for moving from one set to the next can be produced using the *odometer principle*, based on the odometer or mileage gauge in a car. Represent a subset as above by a string of zeros and ones. To construct the next subset in the list, first identify the longest substring of ones at the end of the string. If the string consists entirely of ones, then it is last in the order, and we have finished. Otherwise, this string is preceded by a zero; change the zero to a one, and the ones following it to zeros. For example, for $n = 3$, the odometer principle generates the strings

$$000, 001, 010, 011, 100, 101, 110, 111,$$

which correspond to the subsets

$$\emptyset, \{3\}, \{2\}, \{2,3\}, \{1\}, \{1,3\}, \{1,2\}, \{1,2,3\}$$

of $\{1,2,3\}$.

Notice that the binary strings are in *lexicographic order*, the order in which they would appear in a dictionary, regarding them as words over the alphabet $\{0,1\}$.

For $0 \le k \le n$, the number of k-element subsets of $\{1, \ldots, n\}$ is given by the *binomial coefficient*

$$\binom{n}{k} = \frac{n(n-1) \cdots (n-k+1)}{k(k-1) \cdots 1}.$$

The binomial coefficients are traditionally written in a triangular array where, for $n \ge 0$, the nth row contains the numbers $\binom{n}{0}$, $\binom{n}{1}$, ..., $\binom{n}{n}$. This is usually called *Pascal's triangle*, though, as we will see, it was not invented by Pascal. It begins like this:

$$
\begin{array}{ccccccccc}
 & & & & 1 & & & & \\
 & & & 1 & & 1 & & & \\
 & & 1 & & 2 & & 1 & & \\
 & 1 & & 3 & & 3 & & 1 & \\
1 & & 4 & & 6 & & 4 & & 1
\end{array}
$$

The most important property, and the reason for the name, is the form of the generating function for these numbers (regarded as a sequence indexed by k for fixed n), the *Binomial Theorem*:

$$\sum_{k=0}^{n} \binom{n}{k} x^k = (1+x)^n.$$

Example: permutations   A *permutation* of the set $\{1, \ldots, n\}$ is a rearrangement of the elements of the set, that is, a bijective function from the set to itself. The number of permutations of $\{1, \ldots, n\}$ is the *factorial function* $n!$, defined as usual as the product of the natural numbers from 1 to n. This formula is not so satisfactory, involving an n-fold product. It can be expressed in other ways, as a sum:

$$n! = \sum_{k=0}^{n} (-1)^{n-k} \binom{n}{k} (n-k)^n,$$

or as an integral:

$$n! = \int_0^\infty x^n e^{-x} \, dx.$$

Neither of these is easier to evaluate than the original definition. (We will meet both these formulae later on.)

The recurrence relation for $F(n) = n!$ is

$$F(0) = 1, \qquad F(n) = nF(n-1) \text{ for } n \geq 1.$$

This leads to the same method of evaluation as we saw earlier.

The ordinary generating function for $F(n) = n!$ fails to converge anywhere except at the origin. The exponential generating function is $1/(1-x)$, convergent for $|x| < 1$.

As an example to show that convergence is not necessary for a power series to be useful, let

$$\left(1 + \sum_{n \geq 1} n! x^n \right)^{-1} = 1 - \sum_{n \geq 1} c(n) x^n.$$

Then $c(n)$ is the number of connected permutations on $\{1, \ldots, n\}$. (A permutation π is *connected* if there does not exist k with $1 \leq k \leq n-1$ such that π maps $\{1, \ldots, k\}$ to itself.) This will be proved in the next chapter.

The approximate size of the factorial function is not obvious, as it was for powers of 2. An asymptotic estimate for $n!$ is given by *Stirling's formula*:

$$n! \sim \sqrt{2\pi n} \left(\frac{n}{e}\right)^n.$$

We give the proof later.

It is possible to generate permutations sequentially, or choose a random permutation, by a method similar to that for subsets, using a variable base.

The set of permutations of $\{1, \ldots, n\}$ forms a group under the operation of composition, the *symmetric group* of degree n, denoted by S_n.

Example: derangements A derangement is a permutation with no fixed points. Let $d(n)$ be the number of *derangements* of n.

There is a simple formula for $d(n)$: it is the nearest integer to $n!/e$. (This is one of the oldest formulae in combinatorics, having been proved by de Montmort in 1713.) This is also satisfactory as an asymptotic expression for $d(n)$; we can supplement it with the fact that $|d(n) - n!/e| < 1/(n+1)$ for $n > 0$.

This formula is not very good for calculation, since it requires accurate knowledge of e and operations of real (rather than integer) arithmetic. There are, however, two recurrence relations for $d(n)$; the second, especially, leads to efficient calculation:

$$d(0) = 1, d(1) = 0, \qquad d(n) = (n-1)(d(n-1) + d(n-2)) \text{ for } n \geq 2;$$
$$d(0) = 1, \qquad d(n) = nd(n-1) + (-1)^n \text{ for } n \geq 1.$$

The ordinary generating function for $d(n)$ fails to converge, but the exponential generating function is equal to $\exp(-x)/(1-x)$.

These facts will be proved in Chapter 4.

Since the probability that a random permutation is a derangement is about $1/e$, we can choose a random derangement as follows: repeatedly choose a random permutation until a derangement is obtained. The expected number of choices necessary is about e.

Example: partitions   The *partition number* $p(n)$ is the number of non-increasing sequences of positive integers with sum n. There is no simple formula for $p(n)$. However, quite a bit is known about it:

- The ordinary generating function is

$$\sum_{n \geq 0} p(n)x^n = \prod_{k \geq 1} (1-x^k)^{-1}.$$

- There is a recurrence relation:

$$p(n) = \sum (-1)^{k-1} p(n-k(3k-1)/2),$$

where the sum is over all non-zero values of k, positive and negative, for which $n - k(3k-1)/2 \geq 0$. Thus,

$$p(n) = p(n-1) + p(n-2) - p(n-5) - p(n-7) + p(n-12) + \cdots,$$

where there are about $\sqrt{8n/3}$ terms in the sum.

These facts will be proved in Chapter 4.

The asymptotics of $p(n)$ are rather complicated, and were worked out by Hardy, Littlewood, and Rademacher:

$$p(n) \sim \frac{1}{4n\sqrt{3}} e^{\pi \sqrt{2n/3}}$$

(more precise estimates, including a convergent series representation, exist).

Example: set partitions   The *Bell number* $B(n)$ is the number of partitions of the set $\{1, \ldots, n\}$. Again, no simple formula is known, and the asymptotics are very complicated. There is a recurrence relation,

$$B(n) = \sum_{k=1}^{n} \binom{n-1}{k-1} B(n-k),$$

and the exponential generating function is

$$\sum \frac{B(n)x^n}{n!} = \exp(\exp(x) - 1).$$

Based on the recurrence one can derive a sequential generation algorithm, which calls itself recursively.

1.2 About how many?

As noted in the last section, if F and G are two functions on the natural numbers which do not vanish, we write $F \sim G$ if $F(n)/G(n) \to 1$ as $n \to \infty$. If $F(n)$ is the solution to a counting problem and $G(n)$ is a familiar analytic function, this tells us roughly the size of the collection we are counting. For example, we mentioned already that the number $n!$ of permutations of a set of size n is given approximately by *Stirling's formula*

$$n! \sim \sqrt{2\pi n}\left(\frac{n}{e}\right)^n.$$

Since $n!$ is the product of n numbers each at most n, it is clear that $n! \leq n^n$; the asymptotic formula gives a much more precise estimate.

In Chapter 11 we will also introduce further notation for *asymptotic analysis*, and in that chapter and the next we describe a variety of techniques for proving such estimates.

1.3 How hard is it?

A formula like 2^n (the number of subsets of an n-set) can be evaluated quickly for a given value of n. A more complicated formula with multiple sums and products will take longer to calculate. We could regard a formula which takes more time to evaluate than it would take to generate all the objects and count them as being useless in practice, even if it has theoretical value.

Traditional *computational complexity* theory refers to decision problems, where the answer is just 'yes' or 'no' (for example, 'Does this graph have a Hamiltonian circuit?'). The size of an instance of a problem is measured by the number of bits of data required to specify the problem (for example, $n(n-1)/2$ bits to specify a graph on n vertices). Then the time complexity of a problem is the function f, where $f(n)$ is the maximum number of steps required by a Turing machine to compute the answer for an instance of size n. (A *Turing machine* is a simple theoretical model of a computer which is capable of any theoretically-possible computation.)

To allow for variations in the format of the input data and in the exact specification of a Turing machine, complexity classes are defined with a broad brush: for example, P (or 'polynomial-time') consists of all problems whose time complexity is at most n^c for some constant c. (For more details, see Garey and Johnson, *Computers and Intractability*.)

For counting problems, the answer is a number rather than a single Boolean value (for example, 'How many Hamiltonian circuits does this graph have?'). Complexity theorists have defined the complexity class #P ('number-P') for this purpose.

Even this class is not really appropriate for counting problems of the type we mostly consider. Consider, for example, the question 'How many partitions does an n-set have?' The input data is the integer n, which (if written in base 2) requires only $m = \lceil 1 + \log_2 n \rceil$ bits to specify. The question asks us to calculate the Bell number $B(n)$, which is greater than 2^{n-1} for $n > 2$, and so it takes time exponential in m simply to write down the answer! To get round this difficulty, it is usual to pretend that the size of the input data is actually n rather than $\log n$. (We can imagine that n is given by writing n consecutive 1s on the input tape of the Turing machine, that is, by writing n as a tally rather than in base 2.)

We have seen that computing 2^n (the number of subsets of an n-set) requires at most $\log n$ integer multiplications. But the integers may have as many as n digits, so each multiplication takes about n Turing machine steps. Similarly, the solution to a recurrence relation can be computed in time polynomial in n, provided that each individual computation can be.

On the other hand, a method which involves generating and testing every subset or permutation will take exponentially long, even if the generation and testing can be done efficiently.

A notion of complexity relevant to this situation is the polynomial delay model, which asks that the time required to generate each object should be at most n^c for some fixed c, even if the number of objects to be generated is greater than polynomial.

Of course, it is easy to produce combinatorial problems whose solution grows faster than, say, the exponential of a polynomial. For example, how many intersecting families of subsets of an n-set are there? The total number, for n odd, lies between $2^{2^{n-1}}$ and 2^{2^n}, so that even writing down the answer takes time exponential in n.

We will not consider complexity questions further in this book.

1.4   Exercises

1.1 Construct a bijection between the set of all k-element subsets of $\{1,\ldots,n\}$ containing no two consecutive elements, and the set of all k-element subsets of $\{1,\ldots,n-k+1\}$. Hence show that the number of such subsets is $\dbinom{n-k+1}{k}$.

When the UK National Lottery was introduced in 1994, the draw consisted of choosing six distinct numbers randomly from the set $\{1,\ldots,49\}$. What is the probability that the draw contained no two consecutive numbers?

1.2   (a) In Vancouver in 1984, I saw a Dutch pancake house advertised 'a thousand and one combinations' of toppings. What do you deduce?

   (b) More recently McDonalds offered a meal deal with a choice from eight components of your meal, and advertised '40312 combinations'. What do you deduce?

1.3 Prove the second formula for $n!$ given in the text:

$$n! = \int_0^\infty x^n e^{-x}\, dx.$$

1.4 Let $f(n)$ be the number of partitions of an n-set into parts of size 2.

(a) Prove that

$$f(n) = \begin{cases} 0 & \text{if } n \text{ is odd;} \\ 1 \cdot 3 \cdot 5 \cdots (n-1) & \text{if } n \text{ is even.} \end{cases}$$

(b) Prove that the exponential generating function for the sequence $(f(n))$ is $\exp(x^2)$.

(c) Use Stirling's formula to prove that

$$f(n) \sim \sqrt{2}\left(\frac{2m}{e}\right)^m$$

for $n = 2m$.

1.5 Show that it is possible to generate all subsets of $\{1,\ldots,n\}$ successively in such a way that each subset differs from its predecessor by the addition or removal of precisely one element. (Such a sequence is known as a *Gray code*.)

This picture might help.

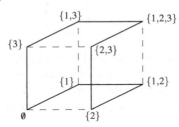

Remark   Gray codes are used in analog-to-digital converters. Since only one digit changes at a time when the input varies continuously, the damage caused by an error in reading the changing digit is minimised.

1.6 Counting can be used to prove structural results, as in the following exercise, which proves a theorem of Mantel: A graph with n vertices and more than $n^2/4$ edges must contain a triangle.

Consider a graph with n vertices, e edges and t triangles. Let x_i be the number of edges containing vertex i.

(a) Show by Inclusion–Exclusion that, if vertices i and j are joined, then at least $x_i + x_j - n$ triangles contain these two vertices.

(b) Hence show that
$$6t \geq \sum_i x_i^2 + \sum_j x_j^2 - 2ne.$$

(c) Use the Cauchy–Schwarz inequality to show that
$$\sum_i x_i^2 \geq 4e^2/n.$$

(d) Deduce that
$$t \geq \tfrac{1}{3}e(4e - n^2)/n.$$

(e) Hence show that, if $e > n^2/4$, the graph contains a triangle.

Remark   The *Cauchy–Schwarz inequality* states that, if $x_1, \ldots, x_n$ and $y_1, \ldots, y_n$ are two sequences of real numbers, then

$$\left(\sum_{i=1}^n x_i y_i \right)^2 \leq \left(\sum_{i=1}^n x_i^2 \right) \left(\sum_{i=1}^n y_i^2 \right).$$

Geometrically, the norm of the vector $\vec{x} = (x_1, \ldots, x_n)$ is

$$\|\vec{x}\| = \left(\sum_{i=1}^n x_i^2 \right)^{1/2}.$$

The Cauchy–Schwarz inequality says that the inner product of two vectors $\vec{x}$ and $\vec{y}$ cannot exceed in modulus the product of the norms of the vectors. Indeed, the ratio $(\vec{x} \cdot \vec{y})/\|x\| \cdot \|y\|$ is equal to the cosine of the angle between $\vec{x}$ and $\vec{y}$.

This is a remarkably useful inequality, in combinatorics as well as other branches of mathematics.

Can you prove the inequality? [**Hint:** Calculate the squared norm of the vector $\vec{x} + \lambda \vec{y}$; this is a quadratic function of λ which can never be negative.]

1.7   (a) Find an iterative method for listing the k-subsets of the natural numbers in *reverse lexicographic order* (that is, ordered by the largest element, and if largest elements are equal then by the second largest, and so on). Thus, for $k = 4$, the list begins

$$0123, 0124, 0134, 0234, 1234, 0125, \ldots$$

Your method should give an algorithm for moving from any subset to the next in the list.

(b) Show that, if $a_1 < \cdots < a_k$, then the position of $\{a_1, \ldots, a_k\}$ in the list is given by

$$\binom{a_1}{1} + \binom{a_2}{2} + \cdots + \binom{a_k}{k}.$$

Can you describe the inverse of this function, which enables us to write down the nth subset in the list?

CHAPTER 2

Formal power series

Formal power series are central to enumerative combinatorics. A formal power series is just an alternative way of representing an infinite sequence of numbers. However, the use of formal power series introduces new techniques, especially from analysis, into the subject, as we will see.

This chapter provides an introduction to formal power series and the operations we can do on them. Examples are taken from elementary combinatorics of subsets, partitions and permutations, and will be discussed in more detail in the next chapter. We also take a first look at the use of analysis for finding (exactly or asymptotically) the coefficients of a formal power series; we return to this in the last two chapters. The exponential, logarithmic and binomial series are of crucial importance; we discuss these, but leave their combinatorial content until the next chapter.

2.1 Fibonacci numbers

Leonardo of Pisa, also known as Fibonacci, published a book in 1202 on the use of Arabic numerals, then only recently introduced to Europe. He contended that they made calculation much easier than existing methods (the use of an abacus, or the clumsy Roman numerals used for records). As an exercise, he gave the following problem:

> At the beginning of the year, I acquire a new-born pair of rabbits. Each pair of rabbits produces a new pair at the age of two months and every subsequent month. How many pairs do I have at the end of the year?

Let F_n be the number of pairs of rabbits at the end of the nth month. Then $F_0 = 1$ (given), and $F_1 = 1$ (since the rabbits do not breed in the first month of life). Also, for $n \geq 2$, we have

$$F_n = F_{n-1} + F_{n-2},$$

since at the end of the nth month, we have all the pairs who were alive a month earlier, and also new pairs produced by all pairs at least two months old (that is, those which were alive two months earlier).

From this recurrence relation, it is easy to calculate that

$$F_2 = 2, F_3 = 3, F_4 = 5, F_5 = 8, \ldots, F_{12} = 233.$$

Fibonacci's book was by no means the first occurrence of these numbers. This representation of the Fibonacci numbers was discussed by Virahanka in the 6th century, in connection with Sanskrit poetry. A vowel in Sanskrit can be long (*guru*) or short (*laghu*). If we assume that a long vowel is twice as long as a short vowel, in how many ways can we make a line of poetry of length n out of long and short vowels? For example, the lines of length 4 are

$$GG, \; LLG, \; LGL, \; GLL, \; LLLL,$$

where G and L denote *guru* and *laghu* respectively.

Let S_n be the number of patterns of long and short vowels in a line of length n (measured in units of the short vowel). We have $S_0 = 1$ and $S_1 = 1$. Moreover, a line of length $n \geq 2$ can be formed either by adding a short vowel to a line of length $n - 1$, or by adding a long vowel to a line of length $n - 2$; all lines are accounted for, and there is no overlap. So $S_n = S_{n-1} + S_{n-2}$ for $n \geq 2$. Now an easy induction shows that $S_n = F_n$ for all $n \geq 0$.

Despite Virahanka's precedence, the numbers F_n are usually referred to as *Fibonacci numbers*.

Can we find a formula for the nth term F_n in this sequence? Doing so directly is not easy. Instead, we construct the power series

$$F(x) = \sum_{n \geq 0} F_n x^n = F_0 + F_1 x + F_2 x^2 + \cdots .$$

What is $(1 - x - x^2)F(x)$? Clearly this product has constant term 1, while the term in x is $(1)(1) + (-1)(1) = 0$. For $n \geq 2$, the term in x^n is $(1)F_n + (-1)F_{n-1} + (-1)F_{n-2} = 0$. In other words, we have

$$(1 - x - x^2)F(x) = 1,$$

so that $F(x) = 1/(1 - x - x^2)$.

This is an explicit formula for the power series; how can we turn it into a formula for the coefficients? In this case, there are many ways to proceed; the most elementary uses partial fractions. Write

$$1 - x - x^2 = (1 - \alpha x)(1 - \beta x),$$

where α and β satisfy $\alpha + \beta = 1$, $\alpha\beta = -1$. Thus α and β are the roots of the quadratic $x^2 - x - 1 = 0$, so we have $\alpha = (1 + \sqrt{5})/2$, $\beta = (1 - \sqrt{5})/2$. The number $-\beta = \alpha - 1$ is the *golden ratio*. Putting $\phi = 1/\alpha = \alpha - 1$, we see that $\phi = (1 - \phi)/\phi$. In other words, if a unit interval is divided into two pieces of lengths ϕ and $1 - \phi$, then the ratio of the smaller to the larger is equal to the ratio of the larger to the whole interval.

Now by the old-fashioned technique of partial fractions, we can write

$$\frac{1}{1 - x - x^2} = \frac{A}{1 - \alpha x} + \frac{B}{1 - \beta x}.$$

We can calculate A and B by clearing denominators and equating powers of x: thus

$$1 = A(1 - \beta x) + B(1 - \alpha x),$$

so that

$$\begin{aligned} A + B &= 1, \\ A\beta + B\alpha &= 0. \end{aligned}$$

The solution of these equations is

$$A = \frac{\alpha}{\alpha - \beta} = \frac{\sqrt{5} + 1}{2\sqrt{5}}, \qquad B = \frac{\beta}{\beta - \alpha} = \frac{\sqrt{5} - 1}{2\sqrt{5}}.$$

Now $1/(1 - \alpha x)$ is the sum of the geometric series

$$\sum_{n \geq 0} (\alpha x)^n = 1 + \alpha x + \alpha^2 x^2 + \cdots,$$

and similarly with β replacing α; so we find that

$$\sum_{n \geq 0} F_n x^n = \sum_{n \geq 0} (A\alpha^n + B\beta^n) x^n.$$

Equating coefficients of powers of x gives

$$F_n = A\alpha^n + B\beta^n = \frac{1}{\sqrt{5}} \left(\frac{1 + \sqrt{5}}{2} \right)^{n+1} - \frac{1}{\sqrt{5}} \left(\frac{1 - \sqrt{5}}{2} \right)^{n+1}.$$

We have succeeded in finding a formula for the Fibonacci numbers. There are several things to note about this formula.

First, it gives us a very good asymptotic estimate. For $\alpha = (1 + \sqrt{5})/2$ is greater than 1, while $\beta = (1 - \sqrt{5})/2$ is between 0 and -1 (in fact, α and β are approximately 1.618 and -0.618 respectively), so, as $n \to \infty$, α^n grows exponentially while β^n decays exponentially to zero. Thus, $F_n \sim A\alpha^n$.

Indeed, the second term in our formula above is always less than $1/2$ in modulus, so we can conclude that F_n is the nearest integer to $(1/\sqrt{5})((1+\sqrt{5})/2)^{n+1}$.

Second, the formula is all but useless for calculation. Evaluating it for $n = 100$ would require doing two binomial expansions, cancelling half the terms, and then knowing $\sqrt{5}$ to sufficient accuracy that we could evaluate the result to within an integer. Using the recurrence relation, as Fibonacci did, is a much more efficient method! We will see an even quicker method in Exercise 2.12, allowing the evaluation of F_n in only $2\log n$ arithmetic operations with integers (though each of these operations will take about n steps, since the integers have this many digits).

Third, it is almost always easier to prove properties of the Fibonacci numbers using the recurrence relation than to use the formula.

Finally, you may feel a bit uneasy about the manipulations used in the argument. In analysis, you learn that these manipulations are justified if the series converge absolutely; we did them with no regard to convergence, and the results we obtain could be used to show that the series do converge. However, it all seems a bit suspicious!

2.1.1 How do the Fibonacci numbers start?

Nobody doubts that the recurrence relation for the Fibonacci numbers is $F_n = F_{n-1} + F_{n-2}$. But there is no general agreement on the 'initial conditions'. There are three fairly common conventions:

(a) $F_0 = 0$, $F_1 = 1$

(b) $F_0 = 1$, $F_1 = 1$

(c) $F_0 = 1$, $F_1 = 2$

Changing from one convention to another simply shifts the sequence by one or two places, and requires a corresponding adjustment in the formula for the nth Fibonacci number.

The second and third conventions are both rather natural if you believe that the Fibonacci numbers count things. If, like Virahanka, you think that F_n is the number of compositions of n made of 1s and 2s, then you will probably use the second convention above; if you think that it is the number of zero-one sequences with no two consecutive ones, then you will use the third convention.

The first convention has some remarkable properties. For a start, it satisfies $\gcd(F_n, F_m) = F_{\gcd(m,n)}$, so that in particular, if m divides n then F_m divides F_n. A consequence is that F_n is prime only if n is prime (apart from one small case: $F_2 = 1$ with this convention, and $F_4 = 3$). Thus, $F_3 = 2$, $F_5 = 5$, $F_7 = 13$, ... There is also the very curious property that $F_{12} = 12^2$.

What convention did Fibonacci himself adopt? He used his problem about rabbits to demonstrate the superiority of the 'arabic' numerals (then new to Europe) for calculation. The Wikipedia article on Fibonacci gives a lovely illustration of a page from Fibonacci's book *Liber Abaci* which has a table of the Fibonacci numbers $F_0, \ldots, F_{12}$, showing clearly that $F_{12} = 377$. (The table taken from this public domain image is shown on the right.) To obtain this value, Fibonacci must have used convention (c) above. This is a little odd, because his famous problem about the breeding rabbits seems to fit more naturally with convention (b), as I did at the start of this chapter, which would give $F_{12} = 233$.

I don't know the solution to this little mystery!

2.2 Formal power series

The answer to a counting problem is a sequence $a_0, a_1, a_2, \ldots$ of non-negative integers, where a_n is the number of objects we are counting based on a set of size n. A glance at the On-line Encyclopedia of Integer Sequences will reveal that very many interesting sequences arise in this way as counting problems.

A sequence is a single mathematical object with infinitely many terms. However, it is convenient to translate it into a different kind of mathematical object. A *formal power series* is a series of the form

$$\sum_{n \geq 0} a_n x^n = a_0 + a_1 x + a_2 x^2 + \cdots.$$

We say that the term a_0 is the *constant term* of the series, and that $a_n x^n$ is the *term of degree n*.

The sequence and the corresponding formal power series convey exactly the same information. But there are two enormous advantages to using power series.

First, we are going to define operations on formal power series: addition, multiplication, substitution, differentiation. These definitions would be unmotivated and hard to remember if we just gave them for sequences; the power series make it clear what the definitions should be.

Second, although none of the arguments we use to produce the power series involve any calculus, it may happen that the power series (regarded as a series in which x is a real or complex variable) converges in some non-trivial interval or region. If so, it defines a complex analytic function, and we can apply all the resources of analysis to it. These let us prove things which would not be available in any other way.

Finally, we note that integers, complex numbers, etc., are not crucial to the definitions which follow. All that we need is that we can add and multiply the entries in the sequences, and a few simple rules are satisfied (the commutative, associative, and identity laws for addition and multiplication, the distributive law, and the existence of additive inverses or negatives of all elements). In other words, the entries should be taken from a *commutative ring with identity*.

We denote the set of all formal power series with coefficients in R, a fixed commutative ring with identity, by $R[[x]]$.

From this point of view, we can regard a *polynomial* with coefficients in R as a formal power series, all of whose terms are zero from some point on. Thus the set $R[x]$ of polynomials can be identified as a subset (indeed, a subring) of $R[[x]]$.

Now we come to the definitions. It is important to note that, even though we write a formal power series as if it were an infinite sum, the definitions below require no infinite processes or taking limits; everything we do will involve only sums and products of finite numbers of terms in the sequences.

Addition   Let $A = \sum a_n x^n$ and $B = \sum b_n x^n$ be formal power series. We define their sum to be

$$A + B = \sum (a_n + b_n) x^n.$$

In terms of the sequences, we simply add corresponding terms.

Scalar multiplication   Similarly, to multiply a formal power series $A = \sum a_n x^n$ by a scalar r, we simply multiply every term by r:

$$rA = \sum (r a_n) x^n.$$

Multiplication   Let $A = \sum a_n x^n$ and $B = \sum b_n x^n$ be formal power series. We define their product to be $AB = C = \sum c_n x^n$, where

$$c_n = \sum_{k=0}^{n} a_k b_{n-k}.$$

Note that calculating the term of degree n requires $n + 1$ multiplications and n additions. The definition is natural: we obtain the term in x^n in the product by multiplying the term in x^k in the first by the term in x^{n-k} in the second for all possible k, and summing the results. The corresponding operation on sequences is often called *convolution*.

Proposition 2.1 *Let R be a commutative ring with identity. Then the set $R[[x]]$ of formal power series over R, with the above operations, is also a commutative ring with identity.*

We will not prove this result, which involves checking all the appropriate axioms for $R[[x]]$, using the fact that they hold in R.

We can extend the definition of multiplication to give the product of a finite number of formal power series in an obvious way.

As usual, we define $A(x)^n$ by repeated multiplication, or by induction:

$$A(x)^0 = 1, \qquad A(x)^{n+1} = A(x)(A(x))^n \text{ for } n \geq 0.$$

Substitution   Let $A = \sum a_n x^n$ and $B = \sum b_n x^n$ be formal power series, and suppose that *the constant term of B is zero*, that is, $b_0 = 0$. Now we define the result of substituting B into A by

$$A(B(x)) = \sum_{n \geq 0} a_n B(x)^n.$$

We have to look more closely to see that this involves only finite operations. Since the constant term of B is zero, an induction argument shows that all the terms in B^m of degrees $0, 1, \ldots, m-1$ are zero. So when we come to evaluate the term of degree m in the infinite sum $\sum a_n B^n$, there is no contribution from any terms B^{m+1} or higher. So this term can be evaluated by calculating the powers of B up to B^m and taking the appropriate linear combination of the terms.

As a very special case, the substitution rule applies with $B(x) = x$, and gives $\sum a_n x^n$ its expected meaning.

Infinite product   Under certain conditions, we can multiply together infinitely many formal power series.

Suppose that $A_1, A_2, \ldots$ are formal power series. We make two assumptions:

(a) all the power series have constant term 1;
(b) there is a function $s(n)$, satisfying $s(n) \to \infty$ as $n \to \infty$, such that the terms of degree $1, 2, \ldots, s(n) - 1$ in A_n are all zero.

Now consider evaluating the infinite product. First note that the constant term should be the (infinite) product of the constant terms in the series. But assumption (a) means that all terms in this product are 1, so we take the constant term in the product also to be 1. More generally, we are allowed to multiply together infinitely many coefficients as long as all but finitely many of them are equal to 1; the product is just the product of the finitely many terms which are not 1.

Now consider calculating the term of degree m in the infinite product. It is built by choosing terms in the factors whose degrees sum to m in all possible ways, multiplying them together, and adding up the result. But assumption (b) shows that there are only finitely many series in which we can choose a non-zero term which is not the constant term, since terms with degree greater than m are not allowed. So the sum we have to evaluate will have only finitely many terms, and the product is well-defined.

Differentiation This operation sounds suspiciously like calculus, but all we take from calculus is the rule for differentiating positive powers of x. We define an operator D on formal power series by the rule that, if $A = \sum a_n x^n$, then $DA = \sum n a_n x^{n-1}$; in other words, the term of degree n in DA is $(n+1)a_{n+1}x^n$. (We don't get a negative power of x since, as expected, the derivative of the constant term is zero.)

As usual, we usually write df/dx instead of Df (assuming that the variable in the formal power series is x). But remember that no calculus is involved here!

Some examples

Example 1 Let $A(x) = \sum x^n$. We claim that $(1-x)A(x) = 1$. The term of degree n in $(1-x)A(x)$ is $(1)(1) + (-1)(1) = 0$ for $n > 0$; the constant term is 1. So the claim is correct.

Naturally, we write $A(x) = (1-x)^{-1}$, in agreement with the usual formula for the sum of a geometric series. More generally,

$$(1-cx)^{-1} = \sum_{n \geq 0} c^n x^n.$$

Also $A(x)^2 = \sum b_n x^n$, where

$$b_n = \sum_{k=0}^{n} 1 \cdot 1 = n+1;$$

in other words, $A(x)^2 = 1 + 2x + 3x^2 + \ldots$.

Example 2 The infinite product

$$\prod_{n \geq 1} (1+x^n)$$

satisfies the hypotheses above, so is defined as a formal power series. Suppose that it is $\sum a_n x^n$. Then the rule above shows that a_n is obtained by selecting factors $(1+x^k)$ with $k \leq n$ for which the exponents add up to n, multiplying the coefficients (these are all 1, so we end up with 1), and summing all these terms.

In other words,

a_n is the number of ways of writing n as a sum of distinct positive integers.

For example, since

$$8 = 7+1 = 6+2 = 5+3 = 5+2+1 = 4+3+1,$$

we see that $a_8 = 6$.

Example 3   Suppose that $A(x) = \sum a_n x^n$ is a formal power series with constant term 1. Then there is a formal power series $B(x)$, also with constant term 1, such that $A(x)B(x) = 1$.

For let $B(x) = \sum b_n x_n$. Then we have $a_0 b_0 = 1$, so $b_0 = 1$. For $n > 1$ we have

$$\sum_{k=0}^{n} a_k b_{n-k} = 0,$$

so

$$b_n = -\sum_{k=1}^{n} a_k b_{n-k}.$$

This is a recurrence relation for b_n; if we know $b_0, b_1, \ldots, b_{n-1}$, then we can calculate b_n. So finding the inverse of a formal power series is equivalent to solving a special sort of recurrence relation; it is linear, and its coefficients are the negatives of the terms in the first series, but unlike Fibonacci's recurrence it is not in general of fixed length.

Let us see a more elaborate example, which was introduced in Chapter 1. Part of the point of this example is that the series that occur converge nowhere except at the origin; so our subject is not simply a branch of analysis.

Let $A(x) = \sum n! x^n$. The coefficient of x^n is the number of permutations of the set $\{1, \ldots, n\}$. Now say that a permutation π is *connected* if there is no number k, with $1 \le k \le n-1$, such that π maps the set $\{1, \ldots, k\}$ into itself (and so also the set $\{k+1, \ldots, n\}$ to itself). Now we claim:

$$A(x)^{-1} = B(x) = 1 - \sum_{n \ge 1} C_n x^n,$$

where C_n is the number of connected permutations of $\{1, \ldots, n\}$.

Proof   Let C_n be the number of connected permutations of $\{1, \ldots, n\}$. Given any permutation π, let k be the smallest number for which π maps the set $\{1, \ldots, k\}$ into itself. Then $k = n$ if and only if π is connected. In general, π is composed of a connected permutation of $\{1, \ldots, k\}$ and an arbitrary permutation of $\{k+1, \ldots, n\}$; so the number of permutations with a given value of k is $C_k(n-k)!$. Summing, we obtain

$$n! = \sum_{k=1}^{n} C_k(n-k)!.$$

Rearranging this equation shows that the numbers C_n satisfy the recurrence relation for the coefficients of $A(x)^{-1}$.

2.3   Variation and generalisation

As we have seen, a formal power series is a single object which is equivalent to an infinite sequence $(a_0, a_1, a_2, \ldots)$ of numbers. We will say that the formal power series $A(x) = \sum a_n x^n$ is the *generating function* of the sequence, or g.f. for short.

This name is not really accurate since, as we have seen, the concept of a formal power series makes sense even when the series does not converge (and so does not define an analytic function). A more accurate term would be *generating power series*; this term is sometimes used, but 'generating function' is more common, so we will stick to that.

Sometimes the series $A(x)$ above is called the *ordinary generating function*, abbreviated to o.g.f. This contrasts with the *exponential generating function* or e.g.f., which is defined by

$$a(x) = \sum_{n \geq 0} \frac{a_n x^n}{n!},$$

in other words the generating function of the sequence with nth term $a_n/n!$. Later in this book we will see many examples, and motivation, for using the exponential generating function in counting problems. The name comes from the fact that the exponential generating function of the all-1 series is the exponential function:

$$e^x = \sum_{n \geq 0} \frac{x^n}{n!}.$$

Note that the exponential generating function will converge in many cases where the ordinary generating function does not.

Another variant is useful when, instead of a single sequence, we have a two-variable array of numbers, say $a_{n,m}$, where n and m are integers and the points (n,m) lie in some appropriate region of the first quadrant. We associate a formal variable with each subscript (say x with n and y with m in this case), and define the *multivariate generating function*

$$A(x,y) = \sum_{n,m} a_{n,m} x^n y^m.$$

We will see an example in the next chapter, for the array of binomial coefficients.

This can be further varied, for example, we might take the ordinary generating function in one variable and the exponential in the other, say

$$\sum_{n,m} \frac{a_{n,m} x^n y^m}{n!}.$$

2.4 Relation with analysis

Our definitions of operations on formal power series were carefully designed so that each coefficient is generated by finitely many additions or multiplications of the coefficients of the constituent series; so no analysis is required for the definitions.

Nonetheless, we often find that our series are actually convergent in some region of the complex plane, maybe even the whole plane. Let us just stop and remind ourselves about convergence of complex power series.

Let $A(x) = \sum a_n x^n$ be a power series with complex coefficients. Then there is a number r (a non-negative real number or ∞) such that

- the series converges if $|x| < r$;
- the series diverges if $|x| > r$.

Here $r = 0$ means that the series diverges for all non-zero x, while $r = \infty$ means that it converges for all x.

The number r is called the *radius of convergence* of the power series. It is obviously unique.

If $r > 0$, then the series converges to a complex analytic function for $|x| < r$. If r is finite, then this function has a singularity at some point x with $|x| = r$.

For example, the geometric series $\sum c^n x^n$ converges for $|x| < 1/|c|$, and the limit is equal to $1/(1 - cx)$ in this domain; it has a singularity (a simple pole) at $x = 1/c$.

Now, in analysis, one learns rules for addition and multiplication of convergent series, and (under extra conditions) for infinite products, substitution, and differentiation of power series. In every case, these definitions agree with the ones we have given. So we have the following result:

Theorem 2.2 *Suppose that a collection of formal power series over $\mathbb{C}$ converge in the domain $|x| < r$, for some positive real number r. Then the formal power series defined by addition, multiplication or differentiation of these series also converge in this domain, and their limits are the sum, product or derivative of the limit functions of the original series.*

There are similar results for infinite products and substitution, but their statements require a little more care. (In our earlier example of an infinite product, the factors are polynomials and so converge everywhere, but it can be shown that the product has radius of convergence 1.)

In other words, we can operate on formal power series with no consideration of convergence; if the series happen to converge, then all our conclusions will be correct for the limit functions of the series.

These theorems allow us to bring all the resources of analysis to bear on combinatorial counting problems. We will see examples of this later. First, here is a very simple example of such a result.

Theorem 2.3 *Let $A(x) = \sum a_n x^n$ be a formal power series with non-negative coefficients. Suppose that the radius of convergence of the series is r, and assume that $r > 0$.*

(a) *If r is finite, and $c = 1/r$, then the nearest singularity to the origin of the limit function is at $x = r$, and for any $\varepsilon > 0$, we have*

- $a_n < (c+\varepsilon)^n$ *for $n \geq n_0(\varepsilon)$;*
- $a_n > (c-\varepsilon)^n$ *for infinitely many values of n.*

(b) *If $r = \infty$, then $a_n = o(c^n)$ for any $c > 1$.*

2.5 Exponential, logarithmic and binomial series

There are three special power series which occur in analysis, which are of great importance in enumeration. First, the exponential series:

$$\exp(x) = \sum_{n \geq 0} \frac{x^n}{n!},$$

or, more generally,

$$\exp(ax) = \sum_{n \geq 0} \frac{a^n x^n}{n!}.$$

Note that we write $\exp(x)$ rather than e^x, to stress that we are thinking of it as a formal power series. Its characteristic property is that it is equal to its derivative:

$$\frac{\mathrm{d}}{\mathrm{d}x} \exp(ax) = a \exp(ax).$$

The exponential series also has a multiplicative property. This will be familiar to you from analysis; we give a combinatorial proof here.

Proposition 2.4

$$\exp(x+y) = \exp(x) \cdot \exp(y).$$

Proof The nth term of the left-hand side is

$$
\begin{aligned}
(x+y)^n/n! &= \sum_{k=0}^{n} \binom{n}{k} x^{n-k} y^k / n! \\
&= \sum_{k=0}^{n} (x^{n-k}/(n-k)!) \cdot (y^k/k!).
\end{aligned}
$$

(In the first line we use the Binomial Theorem to expand $(x+y)^n$; and in getting from the first line to the second we use the formula

$$\binom{n}{k} = \frac{n!}{k!\,(n-k)!}.$$

Now on the right we have exactly the terms of total degree n in the two variables which arise from multiplying the two exponential series on the right. We see that the multiplicative property of the exponential series is equivalent to the Binomial Theorem for positive integer exponents.

The second type of series is the binomial series for arbitrary exponent. First we generalise the definition of binomial coefficients as follows. Let a be an arbitrary complex number, and n a non-negative integer. Define

$$\binom{a}{n} = \frac{a(a-1)\cdots(a-n+1)}{n!}.$$

The numerator and denominator each have n factors, starting at a or n respectively and decreasing by 1 at each step. For $n = 0$, as usual we take the empty product to be 1, and define

$$\binom{a}{0} = 1.$$

If a is a non-negative integer, then the numerator includes a factor 0 for $n > a$, so that the binomial coefficient is zero.

The *binomial series with exponent a* is defined to be the formal power series

$$(1+x)^a = \sum_{n \geq 0} \binom{a}{n} x^n.$$

We see that, if a is a non-negative integer, the series is finite, and agrees with the series $(1+x)^a$ by the Binomial Theorem; so the definition is actually a theorem in this case.

In general, the series converges for $|x| < 1$, and the sum can be shown to agree with the analytic definition of powers with arbitrary exponent.

The *exponent laws* hold:

$$\begin{aligned}
(1+x)^a \cdot (1+x)^b &= (1+x)^{a+b}; \\
((1+x)^a)^b &= (1+x)^{ab}; \\
(1+x)^a(1+y)^a &= (1+x+y+xy)^a.
\end{aligned}$$

In the second and third equations, we use the substitution rule for formal power series to evaluate $((1+x)^a)^b$ or $(1+(x+y+xy))^a$. (In the first case, we let $(1+x)^a = 1+u$, where u is a formal power series with constant term zero; then, as we have seen, the formal power series $(1+u)^b$ is well-defined.)

We also have the familiar property of differentiation:

$$\frac{\mathrm{d}}{\mathrm{d}x}(1+x)^a = a(1+x)^{a-1}.$$

The third important formal power series is the logarithmic series

$$\log(1+x) = \sum_{n \geq 1} \frac{(-1)^{n-1}x^n}{n}.$$

Again this series converges for $|x| < 1$, and the sum function agrees with the analytic definition of the logarithm. Note that we use $\log(1+x)$ rather than $\log x$, since $\log x$ has an essential singularity at the origin and so there is no Taylor series expansion of it there.

We see that

$$(d/dx)\log(1+x) = \sum_{n \geq 0}(-1)^n x^n = (1+x)^{-1}.$$

The logarithm is the inverse of the exponential function:

$$\begin{aligned} \exp(\log(1+x)) &= 1+x, \\ \log(1+(\exp(x)-1)) &= x. \end{aligned}$$

Again, note that $\exp(x) - 1$ has constant term zero and so can be substituted into the logarithmic series.

There is some elaborate combinatorics underlying these two equations. Some of this is discussed in an appendix to the next chapter.

2.6 Exercises

2.1 (a) Let (a_n) be a sequence of integers, and (b_n) the sequence of partial sums of (a_n) (in other words, $b_n = \sum_{i=0}^{n} a_i$). Suppose that the generating function for (a_n) is $A(x)$. Show that the generating function for (b_n) is $A(x)/(1-x)$.

(b) Let (a_n) be a sequence of integers, and let $c_n = na_n$ for all $n \geq 0$. Suppose that the generating function for (a_n) is $A(x)$. Show that the generating function for (c_n) is $x(d/dx)A(x)$. What is the generating function for the sequence $(n^2 a_n)$?

(c) Use the preceding parts of this exercise to find the generating function for the sequence whose nth term is $\sum_{i=1}^{n} i^2$, and hence find a formula for the sum of the first n squares.

2.2 Suppose that a collection of complex power series all define functions analytic in some neighbourhood of the origin, and satisfy some identity there. Are we allowed to conclude that this identity holds between the series regarded as formal power series?

2.3 Suppose that $A(x)$, $B(x)$ and $C(x)$ are the exponential generating functions of sequences (a_n), (b_n) and (c_n) respectively. Show that $A(x)B(x) = C(x)$ if and only if

$$c_n = \sum_{k=0}^{n} \binom{n}{k} a_k b_{n-k},$$

where

$$\binom{n}{k} = \frac{n!}{k!\,(n-k)!}.$$

2.4 Let R be a commutative ring with identity, and let $R[[x]]$ denote the set of formal power series over R.

(a) Prove that, with the operations of addition and multiplication defined in this chapter, $R[[x]]$ is a commutative ring with identity.

(b) Prove that a formal power series $\sum a_n x^n$ is a unit in $R[[x]]$ (that is, has a multiplicative inverse) if and only if its constant term a_0 is a unit in R.

(c) Suppose that R is a field. Show that $R[[x]]$ has a unique maximal ideal I, consisting of all formal power series with zero constant term (and this ideal is generated by x). What is $R[[x]]/I$?

(d) Suppose that R is a field. Describe all the ideals in $R[[x]]$.

2.5 Prove that $(d/dx)\exp(ax) = a\exp(ax)$ and $(d/dx)(1+x)^a = a(1+x)^{a-1}$.

Exercises on Fibonacci numbers

In all these exercises except the last, we adopt the convention that $F_0 = F_1 = 1$.

2.6 Verify the following formula for the sloping diagonals of Pascal's triangle:

$$\sum_{i=0}^{\lfloor n/2 \rfloor} \binom{n-i}{i} = F_n.$$

2.7   (a) Prove that $F_{n+1}F_{n-1} - F_n^2 = (-1)^{n-1}$ for $n \geq 1$.

(b) Prove that $F_{2n} = F_n^2 + F_{n-1}^2$ and $F_{2n+1} = F_n(F_{n+1} + F_{n-1})$ for $n \geq 1$.

2.8   (a) Prove that the number of sequences of length n of zeros and ones which have no two consecutive ones is F_{n+1} for $n \geq 0$.

(b) In how many ways can zeros and ones be arranged in n positions around a circle so that no two ones are consecutive?

2.9 Find a recurrence relation which is satisfied by both sequences of alternate Fibonacci numbers, that is, $(F_0, F_2, F_4, F_6, \ldots)$ and $(F_1, F_3, F_5, \ldots)$.

2.10 Let $G(n)$ be the number computed in the following way. Write n as a composition (an ordered sum) of positive integers in all possible ways. For each expression, multiply the summands together; then add all the products formed in this way. For example,

$$4 = 3+1 = 1+3 = 2+2 = 2+1+1 = 1+2+1 = 1+1+2 = 1+1+1+1,$$

so

$$G(4) = 4+3+3+4+2+2+2+1 = 21.$$

Prove that $G(n) = F_{2n-1}$.

2.11 In the previous exercise, instead of multiplying the summands, multiply factors 2^{d-2} for each summand $d > 2$. Show that the answer is F_{2n-2}.

2.12 Let $A(n)$ be the 2×2 matrix given by

$$A(n) = \begin{pmatrix} F_n & F_{n-1} \\ F_{n-1} & F_{n-2} \end{pmatrix}$$

for $n \geq 2$, and let $A(1) = \begin{pmatrix} 1 & 1 \\ 1 & 0 \end{pmatrix}$.

(a) Prove that $A(n)A(1) = A(n+1)$ for $n \geq 1$.

(b) Deduce that $A(n) = A(1)^n$ for $n \geq 1$.

(c) Deduce the second part of Exercise 2.7.

2.13 Use the result of the preceding exercise to show that the Fibonacci number $F(n)$ can be calculated with only $c\log n$ arithmetic operations, for some constant c.

2.14 By computing the coefficient of x^m on the two sides of the identity

$$\frac{1}{1-(a+b)x+abx^2} = \sum_{n=0}^{\infty} ((a+b)x - abx^2)^n,$$

prove that

$$\frac{a^{m+1}-b^{m+1}}{a-b} = \sum_{k=0}^{\lfloor m/2 \rfloor} \binom{m-k}{k}(-ab)^k(a+b)^{m-2k}.$$

By taking a and b to be the roots of the equation $z^2 - z - 1 = 0$, deduce the equality of two expressions for the Fibonacci numbers.

(I am grateful to Marcio Soares for this exercise.)

2.15 This exercise explores some remarkable properties of the Fibonacci recurrence relation. I learned this (and more) from J. H. Conway; Clark Kimberling also proved some of these results. There are further properties not discussed here!

(a) Show that any positive integer n can be written uniquely in the form

$$n = F_{m_1} + F_{m_2} + \cdots + F_{m_k},$$

where $m_1, \ldots, m_k$ are positive integers (in increasing order) with no two consecutive.

(b) Define the *Fibonacci successor function* σ on the set of positive integers as follows: if n is expressed as in the preceding part of the question, then

$$\sigma(n) = F_{m_1+1} + F_{m_2+1} + \cdots + F_{m_k+1}.$$

Show that $\sigma(\sigma(n)) = \sigma(n) + n$ for any positive integer n.

(c) Construct an array of positive integers as follows:

- The zeroth row consists of the Fibonacci numbers F_n for $n \geq 1$: so this row is $(1, 2, 3, 5, \ldots)$.
- The first element in any row after the zeroth is the smallest positive integer which has not yet occurred in the table (so, for example, the first number in the first row is 4). Subsequent elements in the row are found by applying the Fibonacci successor function (so the first row begins $(4, 7, 11, 18, \ldots)$).

Prove that

- every positive integer occurs exactly once in the table;
- given any sequence $(a_1, a_2, \ldots)$ satisfying the Fibonacci recurrence relation $a_{n+2} = a_{n+1} + a_n$, the terms of the sequence from some point on agree with the entries in some row of the table.

(d) Extrapolate every row of the table backwards two places so that the Fibonacci recurrence holds (so, for example, the zeroth row $(1, 2, 3, \ldots)$ is preceded by $(0, 1)$. Show that the first entry in the new nth row of the table is n.

2.16 For this exercise, we adopt the convention that $F_0 = 0$ and $F_1 = 1$.
   Prove that

(a) $\gcd(F_m, F_n) = F_{\gcd(m,n)}$;

(b) if $m \mid n$, then $F_m \mid F_n$;

(c) if F_n is prime, then either n is prime or $n = 4$.

Is the converse of the last statement true?

CHAPTER 3

Subsets, partitions and permutations

The basic objects of combinatorics are subsets, partitions and permutations. In this chapter, we consider the problem of counting these. The counting functions have two parameters: n, the size of the underlying set; and k, a measure of the object in question (the number of elements of a subset, parts of a partition, or cycles of a permutation respectively). The corresponding counting functions are binomial coefficients and the two kinds of Stirling numbers; we study them and their generating functions in this chapter. We also give a test for unimodality of a finite sequence and apply it to the binomial coefficients and unsigned Stirling numbers of both kinds.

3.1 Subsets

The number of k-element subsets of the set $\{1, \ldots, n\}$ is the *binomial coefficient*

$$\binom{n}{k} = \begin{cases} 0 & \text{if } k < 0 \text{ or } k > n; \\ \dfrac{n(n-1)\cdots(n-k+1)}{k(k-1)\cdots 1} & \text{if } 0 \le k \le n. \end{cases}$$

For, if $0 \le k \le n$, there are $n(n-1)\cdots(n-k)$ ways to choose in order k distinct elements from $\{1, \ldots, n\}$; each k-element subset is obtained from $k!$ such ordered selections. The result for $k < 0$ or $k > n$ is clear.

Remark The notation $\binom{n}{k}$ is an example of *Karamata–Knuth notation*. We will see further examples when we meet the Stirling numbers. It is not universally popular among mathematicians: some people write ^nC_k instead. Probably a notation like $B(n,k)$ would be preferable, but this is unlikely to be generally adopted.

Proposition 3.1 *The recurrence relation for the binomial coefficients is*

$$\binom{n}{0} = \binom{n}{n} = 1, \qquad \binom{n}{k} = \binom{n-1}{k-1} + \binom{n-1}{k} \ for \ 0 < k < n.$$

Proof Partition the k-element subsets into two classes: those containing n (which have the form $\{n\} \cup L$, where L is a $(k-1)$-element subset of $\{1, \ldots, n-1\}$, and so are $\binom{n-1}{k-1}$ in number); and those not containing n (which are k-element subsets of $\{1, \ldots, n-1\}$, and so are $\binom{n-1}{k}$ in number).

The *Binomial Theorem* for natural number exponents n asserts:

Proposition 3.2 $(x+y)^n = \displaystyle\sum_{k=0}^{n} \binom{n}{k} x^{n-k} y^k.$

Proof The proof is straightforward. On the left we have the product

$$(x+y)(x+y)\cdots(x+y) \qquad (n \text{ factors});$$

multiplying this out we get the sum of 2^n terms, each of which is obtained by choosing y from a subset of the factors and x from the remainder. There are $\binom{n}{k}$ subsets of size k, and each contributes a term $x^{n-k} y^k$ to the sum, for $k = 0, \ldots, n$.

Once we have the Binomial Theorem, alternative proofs of various facts about the binomial coefficient become available. Here, for example, is a proof of Proposition 3.1.

Proof We have

$$(x+y)^n = (x+y)^{n-1} \cdot (x+y).$$

The coefficient of $x^{n-k} y^k$ on the left is $\binom{n}{k}$. On the right, this term arises from multiplying the term in $x^{n-k} y^{k-1}$ in $(x+y)^{n-1}$ by y, and also from multiplying the term $x^{n-k-1} y^k$ by x; so the resulting coefficient is $\binom{n-1}{k-1} + \binom{n-1}{k}$, as required.

The Binomial Theorem can be looked at in various ways. From one point of view, it gives the generating function for the binomial coefficients $\binom{n}{k}$ for fixed n:

$$\sum_{k \geq 0} \binom{n}{k} y^k = (1+y)^n.$$

Since the binomial coefficients have two indices, we could ask for a two-variable generating function:

$$\sum_{n\geq 0}\sum_{k\geq 0}\binom{n}{k}x^n y^k = \sum_{n\geq 0}x^n(1+y)^n$$

$$= \frac{1}{1-x(1+y)}.$$

If we expand this in powers of y, we obtain

$$\frac{1}{(1-x)-xy} = \frac{1}{1-x}\cdot\frac{1}{1-(x/(1-x))y}$$

$$= \sum_{k\geq 0}\left(\frac{x^k}{(1-x)^{k+1}}\right)y^k,$$

so that we have the following:

Proposition 3.3 $\displaystyle\sum_{n\geq k}\binom{n}{k}x^n = \frac{x^k}{(1-x)^{k+1}}.$

Our next observation on the Binomial Theorem concerns *Pascal's triangle*, the triangle whose nth row contains the numbers $\binom{n}{k}$ for $0\leq k\leq n$. (Despite the name, this triangle was not invented by Pascal but occurs in earlier Chinese sources. Figure 3.1[1] shows the triangle as given in Zhu Shijie's *Siyuan Yujian*, dated 1303. Zhu's name is sometimes transliterated as Chu Shi-Chieh, and his book as *Ssu Yuan Yü Chien*.) The recurrence relation shows that each entry of the triangle is the sum of the two above it.

At risk of making the triangle asymmetric, we turn it into a matrix $B = (b_{nk})$, where $b_{nk} = \binom{n}{k}$ for $n,k\geq 0$. This infinite matrix is lower triangular, with ones on the diagonal. Now when two lower triangular matrices are multiplied, each term of the product is only a finite sum: the (n,k) entry of BC is $\sum_m b_{nm}c_{mk}$, and this is non-zero only for $k\leq m\leq n$. In particular, we can ask 'What is the inverse of B?'

The *signed matrix of binomial coefficients* is the matrix B^* with (n,k) entry $(-1)^{n-k}\binom{n}{k}$. That is, it is the same as B except that signs of alternate terms are changed in a chessboard pattern. Now:

Proposition 3.4 *The inverse of the matrix B of binomial coefficients is the matrix B^* of signed binomial coefficients.*

Proof We consider the vector space of polynomials (over $\mathbb{R}$). There is a natural basis consisting of the polynomials $1, x, x^2, \ldots$. Now, since

$$(1+x)^n = \sum_k\binom{n}{k}x^k,$$

---

[1]Used under Creative Commons license by-sa 3.0, `https://creativecommons.org/licenses/by-sa/3.0/`

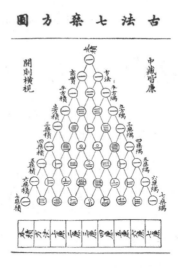

Figure 3.1: Zhu Shijie's triangle

we see that B represents the change of basis to $1, y, y^2, \ldots$, where $y = 1 + x$. Hence the inverse of B represents the basis change in the other direction, given by $x = y - 1$. Since

$$(y-1)^n = \sum_k (-1)^{n-k} \binom{n}{k} y^k,$$

the matrix of this basis change is B^*.

There is a huge literature on 'binomial coefficient identities'. Some of these are given as exercises to this chapter. Here is just one, for which we provide two different proofs.

Proposition 3.5 (Vandermonde convolution) *Let n, m, k be natural numbers. Then*

$$\sum \binom{m}{i} \binom{n}{k-i} = \binom{n+m}{k}.$$

Proof The sum is over the finite number of values of i for which both the binomial coefficients on the right are non-zero; that is, $0 \le i \le m$ and $0 \le k - i \le n$.

Consider the coefficient of x^k in the expansion of $(1+x)^{m+n}$. This is clearly $\binom{m+n}{k}$. On the other hand, we can write

$$(1+x)^{m+n} = (1+x)^m (1+x)^n,$$

so the coefficient of x^k is obtained by summing over i the product of the coefficient of x^i in $(1+x)^m$ and the coefficient of x^{k-i} in $(1+x)^n$, giving the required formula.

Here is a 'combinatorial' proof. The right-hand side counts the number of k-element subsets of a set of $m+n$ objects. Suppose that m of these objects are coloured red and the remaining n are blue. The number of ways of choosing k objects of which i are red and the rest are blue is $\binom{m}{i}\binom{n}{k-i}$; summing gives the result.

The other aspect of the Binomial Theorem is its generalisation to arbitrary real exponents (due to Isaac Newton). This depends on a revised definition of the binomial coefficients.

Let a be an arbitrary real (or complex) number, and k a non-negative integer. Define

$$\binom{a}{k} = \frac{a(a-1)\cdots(a-k+1)}{k!}.$$

Note that this agrees with the previous definition in the case when n is a non-negative integer, since if $k > n$ then one of the factors in the numerator is zero. We do not define this version of the binomial coefficients if k is not a natural number.

Now the *Binomial Theorem* asserts that, for any real number a, we have

$$(1+x)^a = \sum_{k \geq 0} \binom{a}{k} x^k. \tag{3.1}$$

Is this a theorem or a definition? If we regard it as an equation connecting real functions (where the left-hand side is defined by

$$(1+x)^a = \exp(a\log(1+x)), \tag{3.2}$$

and the series on the right-hand side is convergent for $|x| < 1$), it is a theorem, and was understood by Newton in this form. As an equation connecting formal power series, we may follow the same approach, or we may instead choose to regard (3.1) as the definition and (3.2) as the theorem, according to taste. Whichever approach we take, we need to know that the laws of exponents hold:

$$\begin{aligned}
(1+x)^a \cdot (1+y)^a &= (1+(x+y+xy))^a, \\
(1+x)^{a+b} &= (1+x)^a \cdot (1+x)^b, \\
(1+x)^{ab} &= ((1+x)^a)^b.
\end{aligned}$$

If (3.1) is our definition, these verifications will reduce to identities between binomial coefficients; if (3.2) is the definition, they depend on properties of the power series for exp and log, defined as in the last chapter.

3.1.1 Sampling

I have n distinguishable objects in a hat. In how many ways can I choose k of them?

The answer depends on how the sampling is done. Two questions must be resolved:

(a) Is the sampling *with replacement* (after each item is chosen, I note it down and return it to the hat), or *without replacement* (an item once chosen is put to one side and not used again)?

(b) Does the order in which items are chosen matter or not?

Sampling provides a link between combinatorics and elementary probability theory, and also results in the informal name 'permutations and combinations' for the subject (permutations and combinations being ordered or unordered samples).

Theorem 3.6 *The number of ways of sampling k objects from a set of n is given in the following table, where $(n)_k = n(n-1)\cdots(n-k+1)$:*

	Order significant	Order not significant
With replacement	n^k	$\binom{n+k-1}{k}$
Without replacement	$(n)_k$	$\binom{n}{k}$

Remark The number $\binom{n+k-1}{k}$ of samples with replacement and with order not significant can be written as a negative binomial coefficient:

$$\binom{n+k-1}{k} = (-1)^k \binom{-n}{k}.$$

Proof Three of the entries of the table are straightforward. The one that requires proof is the number of samples with replacement and with order not significant. We calculate this by transforming the problem into another.

Step 1 The number of samples of k objects from n, with replacement and with order not significant, is equal to the number of n-tuples $(a_1, \ldots, a_n)$ of non-negative integers with sum k.

We code a sample by the n-tuple whose ith entry a_i is the number of occurrences of the ith object in the sample. Clearly $a_i \geq 0$ and $a_1 + \cdots + a_n = k$. The bijection reverses.

Step 2 The number of n-tuples of non-negative integers with sum k is $\binom{n+k-1}{k}$.

To show this we construct a bijection between the set of n-tuples described and the set of $(n-1)$-element subsets of a set of size $n+k-1$: note that

$$\binom{n+k-1}{n-1} = \binom{n+k-1}{k}.$$

Take $n+k-1$ boxes in a row. Select $n-1$ of them and place barriers in the chosen boxes. Now let a_1 be the number of empty boxes before the first barrier; a_i the number of empty boxes between the $(i-1)$st and ith barriers, for $2 \le i \le n-1$; and a_n the number of empty boxes after the $(n-1)$st barrier. Clearly $a_1, \ldots, a_n$ are non-negative and sum to k. Conversely, any such n-tuple can be encoded in this way: place a barrier after a_1 empty boxes; the next barrier after a_2 further empty boxes; and so on.

So the theorem is proved.

There is a bijective proof of the last part which gives further insight. We have to show that, for unordered samples, the number of samples of k from n with repetitions allowed is equal to the number of samples of k from $n+k-1$ with repetitions not allowed.

Take a sample of k numbers from $\{1, \ldots, n\}$, say $a_1, \ldots, a_k$. Since the sample is unordered, we can order it any way we choose; suppose that $a_1 \le a_2 \le \cdots \le a_k$. Now for $i = 1, \ldots, k$, let $b_i = a_i + i - 1$. Then it is clear that $b_1, \ldots, b_k$ are all chosen from $\{1, \ldots, n+k-1\}$. Also, $b_1, \ldots, b_k$ are all distinct: for even if $a_i = a_{i+1}$, say, the corresponding bs are unequal since they have different numbers added to them. [You should prove the assertion formally.] Moreover, the correspondence can be reversed: if $b_1, \ldots, b_k$ are distinct numbers from $\{1, \ldots, n+k-1\}$, arranged in strictly increasing order, and $a_i = b_i - i + 1$ for $i = 1, \ldots, k$, then $a_1, \ldots, a_k$ are chosen from $\{1, \ldots, n\}$.

Since the sets of samples are bijective, the number of unordered samples from $\{1, \ldots, n\}$ with repetitions allowed is $\binom{n+k-1}{k}$.

Remark This result can be formulated in various ways.

(a) In how many ways can I distribute k identical sweets among n children? If the ith child gets a_i sweets, then $a_i \ge 0$ and $a_1 + \cdots + a_n = k$; so the number of ways is $\binom{n+k-1}{k}$.

(b) In how many ways can I distribute the sweets if it is required that each child should have at least one sweet? First give each child one sweet, and distribute the remaining $k-n$ sweets; this can be done in $\binom{k-1}{k-n} = \binom{k-1}{n-1}$ ways.

(c) Given n commuting indeterminates $x_1, \ldots, x_n$, how many monomials of total degree k are there? Such a monomial has the form $x_1^{a_1} \cdots x_n^{a_n}$, where $a_i \geq 0$ and $a_1 + \cdots + a_n = k$; so the number is $\binom{n+k-1}{k}$.

Interchanging the roles of n and k, the result in (b) asserts that the number of *compositions* of n with k parts (that is, expressions for n as an ordered sum of k positive integers) is $\binom{n-1}{k-1}$. Summing over k, the total number of compositions of n is

$$\sum_{k=1}^{n} \binom{n-1}{k-1} = 2^{n-1}$$

(compare the introductory example in Chapter 4).

3.1.2 Digression: Computing binomial coefficients

The formula

$$\binom{n}{k} = \frac{n!}{k!\,(n-k)!}$$

is compact to write down but is not good for calculation. If k is much smaller than n, then we have to compute the huge number $n!$ and divide it by the huge number $(n-k)!$ to come up with a much smaller answer. For example,

$\binom{100}{3} = \frac{93326215443944152681699238856266700490715968264381621468592963895217599993229915608941463976156518286253697920827223758251185210916866\ldots}{6 \times 961927596824821198533284259495636987123438139191729761581044773193337456124818754988058791755890726512612841896796781676470678327\ldots}$

You see it won't fit on the page, even at the smallest type size. The answer is 161700.

How can we find this more efficiently?

If k is small compared to n, the formula

$$\binom{n}{k} = \frac{n(n-1)\cdots(n-k+1)}{k!}$$

uses only $2(k-1)$ multiplications and one division. Even better is to do the multiplications and divisions alternately:

$$\binom{n}{k} = \binom{n}{k-1} \cdot \frac{n-k+1}{k}.$$

This has $k-1$ multiplications and the same number of divisions, but the numbers being divided by are at most k and the numbers in the sum never grow too large. Moreover, in the sequence

$$\binom{n}{k} = \frac{n}{1} \times \frac{n-1}{2} \times \frac{n-2}{3} \times \cdots$$

doing multiplications and divisions alternately, after the lth division we have $\binom{n}{l}$, so we get the intermediate ones too.

If you have to work out all the binomial coefficients $\binom{n}{k}$ for fixed n, you can use Pascal's triangle. Only about $n^2/2$ additions are required, and additions are quicker to perform than multiplications and divisions. Moreover, the number can be roughly halved by using the symmetry of Pascal's triangle:

$$\binom{n}{k} = \binom{n}{n-k}.$$

Binomial coefficients can be estimated by using Stirling's formula. (See Exercise 3.8, for example.)

The *Central Limit Theorem* from probability theory can also be used to get estimates for binomial coefficients. Suppose that a fair coin is tossed n times. Then the probability of obtaining k heads is equal to $\binom{n}{k}/2^n$. Now the number of heads is a binomial random variable X; so we have

$$\mathbb{P}(X = k) = \binom{n}{k}\Big/2^n. \tag{3.3}$$

According to the Central Limit Theorem, if n is large then X is approximated by a normal random variable Y with the same expected value $n/2$ and variance $n/4$. The probability density function of Y is given by

$$\mathbb{P}(Y = y) = \frac{1}{\sqrt{\pi n/2}} e^{-2(k-n/2)^2/n}. \tag{3.4}$$

If $k = n/2 + c\sqrt{n}$ and $n \to \infty$, then a precise statement of the Central Limit Theorem shows that (3.4) gives an asymptotic formula for (3.3). In particular, when $k = n/2$, we obtain the result of Exercise 3.8.

3.2 Partitions

In this section, we see for the first time a distinction between 'labelled' and 'unlabelled' structures, which will be discussed in more detail in Chapters 7 and 10. A labelled structure of a certain type is simply a structure on the set $\{1, 2, \ldots, n\}$, for some n. An unlabelled structure is an equivalence class of labelled structures, where two structures are equivalent if one can be obtained from the other by some permutation of the set $\{1, 2, \ldots, n\}$.

For a rather trivial example, consider the case where our 'structures' are subsets of the set in question. The number of labelled structures is simply the number of subsets of $\{1, \ldots, n\}$, which is 2^n. Now two subsets are equivalent in the above sense if and only if they have the same cardinality. So the number of unlabelled subsets of an n-set is the number of possible cardinalities: the possibilities are $0, 1, 2, \ldots, n$, so the number is $n + 1$.

The *Bell number $B(n)$* is the number of partitions of the set $\{1, \ldots, n\}$. This is a count of labelled structures. Two partitions are equivalent if and only if the lists of sizes of the parts, arranged in non-increasing order, are the same in the two cases. (For if this holds, we may find a permutation mapping the ith part of one partition to the ith part of the other, for all i.) So the 'unlabelled' counting number is $p(n)$, the *partition number*, which is the number of partitions of the number n (that is, lists in non-increasing order of positive integers with sum n). Thus, given any set partition, the list of sizes of its parts is a number partition; and two set partitions are equivalent under relabelling the elements of the underlying set (that is, under permutations of $\{1, \ldots, n\}$) if and only if the corresponding number partitions are equal.

3.2.1 Set partitions

The *Stirling numbers of the second kind*, denoted by $S(n, k)$, are defined by the rule that $S(n, k)$ is the number of partitions of $\{1, \ldots, n\}$ into k parts if $1 \leq n \leq k$, and zero otherwise. Clearly we have

$$\sum_{k=1}^{n} S(n, k) = B(n),$$

where the Bell number $B(n)$ is the total number of partitions of $\{1, \ldots, n\}$.

By analogy with the binomial coefficients, Karamata proposed the notation $\begin{Bmatrix} n \\ k \end{Bmatrix}$ for $S(n, k)$. This approach was championed by Knuth, and is sometimes referred to as *Karamata–Knuth notation*.

Proposition 3.7 *The recurrence relation for the Stirling numbers is*

$$S(n, 1) = S(n, n) = 1, \qquad S(n, k) = S(n-1, k-1) + kS(n-1, k) \text{ for } 1 < k < n.$$

Proof We split the partitions into two classes: those for which $\{n\}$ is a single part (obtained by adjoining this part to a partition of $\{1, \ldots, n-1\}$ into $k-1$ parts), and the remainder (obtained by taking a partition of $\{1, \ldots, n-1\}$ into k parts, selecting one part, and adding n to it). There are $S(n-1, k-1)$ partitions in the first class, and $kS(n-1, k)$ in the second.

Proposition 3.8 *(a) The Stirling numbers satisfy the recurrence*

$$S(n, k) = \sum_{i=1}^{n-1} \binom{n-1}{i-1} S(n-i, k-1).$$

(b) The Bell numbers satisfy the recurrence

$$B(n) = \sum_{i=1}^{n} \binom{n-1}{i-1} B(n-i).$$

Proof Consider the part containing n of an arbitrary partition with k parts; suppose that it has cardinality i. Then there are $\binom{n-1}{i-1}$ choices for the remaining $i-1$ elements in this part, and $S(n-i, k-1)$ partitions of the remaining $n-i$ elements into $k-1$ parts. This proves (a); the proof of (b) is almost identical.

The Stirling numbers also have the following property. Let $(x)_k$ denote the polynomial $x(x-1)\cdots(x-k+1)$.

Proposition 3.9 $x^n = \sum_{k=1}^{n} S(n,k)(x)_k.$

Proof We prove this first when x is a positive integer. We take a set X with x elements, and count the number of n-tuples of elements of x. The total number is of course x^n. We now count them another way. Given an n-tuple $(x_1,\ldots,x_n)$, we define an equivalence relation on $\{1,\ldots,n\}$ by $i \equiv j$ if and only if $x_i = x_j$. If this relation has k different classes, then there are k distinct elements among $x_1,\ldots,x_n$, say $y_1,\ldots,y_k$ (listed in order). The choice of the partition and the k-tuple $(y_1,\ldots,y_k)$ uniquely determines $(x_1,\ldots,x_n)$. So the number of n-tuples is given by the right-hand expression also.

Now this equation between two polynomials of degree n holds for any positive integer x, so it must be a polynomial identity.

We can find a generating function in terms of n:

Theorem 3.10 *For $k \geq 1$,*

$$\sum_{n\geq 1} S(n,k)y^n = \frac{y^k}{(1-y)(1-2y)\cdots(1-ky)}.$$

Proof We use induction on k. For $k=1$, we have $S(n,k) = 1$ for all n; so

$$\sum_{n\geq 1} S(n,1)y^n = y + y^2 + y^3 + \cdots = \frac{y}{1-y},$$

so the formula is true for $k=1$.

Now suppose that it is true for $k-1$. Let $F_k(y) = \sum_n S(n,k)y^n$, so the induction hypothesis is $F_{k-1}(y) = y^{k-1}/(1-y)\cdots(1-(k-1)y)$.

Remember that we have the recurrence

$$S(n,k) = S(n-1,k-1) + kS(n-1,k).$$

Multiplying this equation by y^n and summing over n, we have

$$\sum_{n\geq 1} S(n,k)y^n - ky \sum_{n\geq 1} S(n-1,k)y^{n-1} = y \sum_{n\geq 1} S(n-1,k-1)y^{n-1}.$$

Now in the terms involving $S(n-1,*)$, we first note that we can start the summation at $n=2$ since the terms for $n=1$ are zero; then use a new variable $m = n-1$; and finally replace the variable m by n. We get

$$F_k(y) - kyF_k(y) = yF_{k-1}(y),$$

so that

$$F_k(y) = \frac{y}{1-ky} F_{k-1}(y) = \frac{y^k}{(1-y)\cdots(1-ky)}.$$

Stirling numbers are involved in the substitution of $\exp(x) - 1$ for x in formal power series. The result depends on the following lemma:

Lemma 3.11

$$\sum_{n\geq k} \frac{S(n,k)x^n}{n!} = \frac{(\exp(x)-1)^k}{k!}.$$

Proof The proof is by induction on k, the result being true when $k=1$ since $S(n,1) = 1$. Suppose that it holds when $k = l-1$. Then (setting $S(n,k) = 0$ if $n < k$) we have

$$\begin{aligned}
\frac{(\exp(x)-1)^l}{l!} &= \frac{1}{l} \cdot (\exp(x)-1) \cdot \frac{(\exp(x)-1)^{l-1}}{(l-1)!} \\
&= \frac{1}{l} \left(\sum_{n\geq 1} \frac{x^n}{n!} \right) \cdot \left(\sum_{n\geq 1} \frac{S(n,l-1)x^n}{n!} \right).
\end{aligned}$$

The coefficient of $x^n/n!$ here is

$$\begin{aligned}
\frac{n!}{l} \sum_{i=1}^{n-1} \frac{1}{i!} \cdot \frac{S(n-i,l-1)}{(n-i)!} &= \frac{1}{l} \sum_{i=1}^{n-1} \binom{n}{i} S(n-i,l-1) \\
&= \frac{1}{l} (S(n+1,l) - S(n,l-1)),
\end{aligned}$$

using the recurrence relation of Proposition 3.8(a). Finally, the recurrence relation of Proposition 3.7 shows that this is $S(n,l)$, as required.

Proposition 3.12 *Let $(a_0, a_1, \ldots)$ and $(b_0, b_1, \ldots)$ be two sequences of numbers, with exponential generating functions $A(x)$ and $B(x)$ respectively. Then the following two conditions are equivalent:*

(a) $b_0 = a_0$ and $b_n = \sum_{k=1}^{n} S(n,k)a_k$ for $n \geq 1$;

(b) $B(x) = A(\exp(x) - 1)$.

Proof Suppose that (a) holds. Without loss of generality we may assume that $a_0 = b_0 = 0$. Then

$$
\begin{aligned}
B(x) &= \sum_{n \geq 1} \frac{b_n x^n}{n!} \\
&= \sum_{n \geq 1} \frac{x^n}{n!} \sum_{k=1}^{n} S(n,k)a_k \\
&= \sum_{k \geq 1} a_k \sum_{n \geq k} \frac{S(n,k)x^n}{n!} \\
&= \sum_{k \geq 1} \frac{a_k(\exp(x) - 1)^k}{k!} \\
&= A(\exp(x) - 1),
\end{aligned}
$$

by Lemma 3.11.

The converse is proved by reversing the argument.

Corollary 3.13 *The exponential generating function for the Bell numbers is*

$$
\sum_{n \geq 0} \frac{B(n)x^n}{n!} = \exp(\exp(x) - 1).
$$

Proof Apply Proposition 3.12 to the sequence with $a_n = 1$ for all n; or sum the equation of Lemma 3.11 over k.

3.2.2 Number partitions

The partition number $p(n)$ is the number of partitions of an n-set, up to permutations of the set.

The key to evaluating $p(n)$ is its generating function:

$$
\sum_{n \geq 0} p(n)x^n = \left(\prod_{k \geq 1} 1 - x^k \right)^{-1}.
$$

For $(1 - x^k)^{-1} = 1 + x^k + x^{2k} + \cdots$. Thus a term in x^n in the product, with coefficient 1, arises from every expression $n = \sum c_k k$, where the c_k are non-negative integers, all but finitely many equal to zero. This number is $p(n)$, since we can regard $n = \sum c_k k$ as an alternative expression for a partition of n.

We will use this in the next chapter to give a recurrence relation for $p(n)$.

3.3 Permutations

A permutation of $\{1, \ldots, n\}$ is a bijective function from this set to itself.

In the nineteenth century, a more logical terminology was used. Such a function was called a substitution, while a permutation was a sequence $(a_1, a_2, \ldots, a_n)$ containing each element of the set precisely once. Since there is a natural ordering of $\{1, 2, \ldots, n\}$, there is a one-to-one correspondence between 'permutations' and 'substitutions': the sequence $(a_1, a_2, \ldots, a_n)$ corresponds to the function $\pi : i \mapsto a_i$, for $i = 1, \ldots, n$.

The correspondence between permutations and total orderings of an n-set has profound consequences for a number of enumeration problems. For now we return to the usage 'permutation = bijective function'. We refer to the sequence $(a_1, \ldots, a_n)$ as the *passive form* of the permutation π in the last paragraph; the function is the *active form* of the permutation.

Following the conventions of algebra, we write a permutation on the right of its argument, so that $i\pi$ is the image of i under the permutation π (that is, the ith term of the passive form of π).

The set of permutations of $\{1, \ldots, n\}$, with the operation of composition, is a group, called the *symmetric group* S_n. Products, identity, and inverses of permutations always refer to the operations in this group.

3.3.1 Cycle structure

In order to treat permutations in a similar way to subsets and partitions, we need a parameter to take the place of the cardinality of a subset, or the number of parts of a partition. For this we must first explain the cycle structure and parity (or sign) of a permutation.

A *cycle* is a permutation which, for some elements $a_1, \ldots, a_r$, maps

$$a_1 \mapsto a_2 \mapsto \cdots \mapsto a_r \mapsto a_1$$

and fixes all other points (if any). A *transposition* is a cycle of length $r = 2$.

Proposition 3.14 *Any permutation of $\{1, \ldots, n\}$ can be written as a product of a pairwise disjoint collection of cycles, uniquely up to the order of the factors and the choices of the starting points of the cycles.*

Proof Take any point $a \in \{1, \ldots, n\}$, and map it repeatedly until it returns to a value previously visited. This point must be a, since any other point will already have a pre-image, and we have a cycle. If not all points have been covered, choose another point and repeat the procedure, producing a cycle disjoint from the first (again, no overlap is possible since all points in previously found cycles already have pre-images). Repeat until all points are included.

This is is called the *cycle decomposition* of the permutation. This provides us with a compact notation for permutations. For example, the permutation which maps $x \mapsto 3x$ (mod 10), acting on $\{1, \ldots, 9\}$, can be written in cycle notation as $(1,3,9,7)(2,6,8,4)(5)$. (Points fixed by the permutation constitute cycles of length 1; they are commonly ignored in cycle notation, but for the considerations here, it is important to include them!)

The *parity* of a permutation π of $\{1, \ldots, n\}$ is defined as the parity of $n - k$, where k is the number of cycles in the cycle decomposition of π (including fixed points). The *sign* of π is $(-1)^{n-k}$, with n and k as above.

Parity and sign have various important algebraic properties. For example, it is easy to see that every permutation can be written as a product of transpositions. For using the cycle decomposition, it suffices to show that a cycle is such a product; we have

$$(a_1, a_2, \ldots, a_r) = (a_1, a_2)(a_1, a_3) \cdots (a_1, a_r).$$

Proposition 3.15 (a) *The parity of π is equal to the parity of the number of factors in any expression for π as a product of transpositions.*

  (b) *Parity is a homomorphism from the symmetric group S_n to the group $\mathbb{Z}/(2)$ of integers mod 2, and hence sign is a homomorphism to the multiplicative group $\{\pm 1\}$.*

  (c) *For $n > 1$, these homomorphisms are onto; their kernel (the set of permutations of even parity, or of sign $+1$) is a normal subgroup of index 2 in S_n, called the* alternating group A_n.

Proof The proof of the first part will be indicated; the rest of the proposition is an exercise. Let π be a permutation, written as a product of disjoint cycles. Show that, for any transposition $\tau = (a, b)$,

- if a and b lie in distinct cycles C_1 and C_2 of π, these two are stitched together into a single cycle of $\pi\tau$;

- conversely, if a and b lie in the same cycle of π, this cycle is cut apart into two cycles of $\pi\tau$.

So the number of cycles increases or decreases by 1, and the parity changes.

Now the empty product of transpositions is the identity permutation, which has n cycles, and so even parity; every transposition changes the parity, so a product of l transpositions has the same parity as l.

3.3.2 Unlabelled permutations

As for partitions, we can consider unlabelled or labelled permutations, that is, permutations of an n-set or equivalence classes of permutations. We dispose of unlabelled permutations first.

First we must explain what 'unlabelled permutations' are. Two permutations π_1 and π_2 of $\{1,\ldots,n\}$ are equivalent if there is a bijection σ of $\{1,\ldots,n\}$ (that is, a permutation!) such that, for all $i \in \{1,\ldots,n\}$, we have

$$(i\sigma)\pi_2 = j\sigma \quad \text{if and only if} \quad i\pi_1 = j,$$

in other words, $i\sigma\pi_2 = i\pi_1\sigma$ for all i, so that $\pi_2 = \sigma^{-1}\pi_1\sigma$. Thus, this equivalence relation is the algebraic relation of *conjugacy* in the symmetric group; the unlabelled permutations are conjugacy classes of S_n.

Now the following property holds:

Proposition 3.16 *Two permutations are conjugate in the symmetric group if and only if the lists of cycle lengths of the two permutations (written in non-increasing order) are equal.*

Thus equivalence classes of permutations correspond to partitions of the integer n. This means that the enumeration theory for 'unlabelled permutations' is the same as that for 'unlabelled partitions', discussed in the last section.

3.3.3 Labelled permutations

The number of permutations of $\{1,\ldots,n\}$ is given by the factorial function $n!$. As in the case of subsets and partitions, we divide the permutations into classes involving an auxiliary parameter k. This parameter will be the number of cycles in the cycle decomposition. However, it is traditional to record the parity as well.

The *Stirling numbers of the first kind* are defined by the rule that $s(n,k)$ is $(-1)^{n-k}$ times the number of permutations of $\{1,\ldots,n\}$ having k cycles. (Note that $(-1)^{n-k}$ is the common sign of these permutations.) Sometimes the number of such permutations is referred to as the *unsigned Stirling number*, denoted by $u(n,k)$.

Clearly we have

$$\sum_{k=1}^{n} u(n,k) = n!\,.$$

Slightly less obviously,

$$\sum_{k=1}^{n} s(n,k) = 0$$

for $n > 1$. The algebraic proof of this depends on the fact that sign is a homomorphism to $\{\pm 1\}$, so that the two values are taken equally often. We saw an algebraic proof of this above. Here is a combinatorial translation. Composition with a fixed transposition τ changes the sign of a permutation, so this map is a bijection between the set of permutations with sign $+1$ and the set of those with sign -1.

There is a Karamata–Knuth notation for unsigned Stirling numbers of the first kind, analogous to that for binomial coefficients and Stirling numbers of the second kind: namely, $\begin{bmatrix} n \\ k \end{bmatrix}$ for $u(n,k)$.

Proposition 3.17 *The recurrence relation for the Stirling numbers is*

$$s(n,1) = (-1)^{n-1}(n-1)!, \quad s(n,n) = 1,$$
$$s(n,k) = s(n-1,k-1) - (n-1)s(n-1,k) \text{ for } 1 < k < n.$$

Proof We split the permutations into two classes: those for which (n) is a single part (obtained by adjoining this cycle to a permutation of $\{1, \ldots, n-1\}$ with $k-1$ cycles), and the remainder (obtained by taking a permutation of $\{1, \ldots, n-1\}$ with k cycles and interpolating n at some position in one of the cycles). The second construction, but not the first, changes the sign of the permutations.

To see that there are $(n-1)!$ permutations with a single cycle, note that if we choose to start the cycle with 1 then the remaining $n-1$ elements can be written into the cycle in any order.

Note that, if we instead define $s(n,0)$ and $s(n,n+1)$ to be equal to 0 for $n \geq 1$, then the recurrence holds also for $k = 1$ and $k = n$. We use this below.

The generating function is given by the following result:

Proposition 3.18 $\displaystyle\sum_{k=1}^{n} s(n,k)x^k = (x)_n.$

Proof The result is clear for $n = 1$. Suppose that it holds for $n = m-1$.

$$\sum_{k=1}^{m} s(m,k)x^k = \sum_{k=1}^{m} s(m-1,k-1)x^k - \sum_{k=1}^{m} (m-1)s(m-1,k)x^k$$
$$= (x-m+1)(x)_{m-1}$$
$$= (x)_m.$$

Note that substituting $x = 1$ into this equation shows that $\sum_k s(n,k) = 0$ for $n \geq 2$.

Corollary 3.19 *The triangular matrices S_1 and S_2 whose entries are the Stirling numbers of the first and second kinds are inverses of each other.*

Proof Propositions 3.9 and 3.18 show that S_1 and S_2 are the transition matrices between the bases $(x^n : n \geq 1)$ and $((x)_n \, n \geq 1)$ of the space of real polynomials with constant term zero.

Proposition 3.20 *Let $(a_0, a_1, \ldots)$ and $(b_0, b_1, \ldots)$ be two sequences of numbers, with exponential generating functions $A(x)$ and $B(x)$ respectively. Then the following two conditions are equivalent:*

(a) $b_0 = a_0$ and $b_n = \sum_{k=1}^{n} s(n,k) a_k$ for $n \geq 1$;

(b) $B(x) = A(\log(1+x))$.

Proof This is the 'inverse' of Proposition 3.12.

We have counted permutations by number of cycles. A more refined count is by the list of cycle lengths.

Let $c_k(\pi)$ be the number of k-cycles in the cycle decomposition of π.

Proposition 3.21 *The size of the conjugacy class of π in S_n is*

$$\frac{n!}{\prod_k k^{c_k(\pi)} c_k(\pi)!}.$$

Proof Write out the pattern for the cycle structure of a permutation with $c_k(\pi)$ cycles of length k for all k, leaving blank the entries in the cycles. There are $n!$ ways of entering the numbers $1, \ldots, n$ in the pattern. However, each cycle of length k can be written in k different ways, since the cycle can start at any point; and the cycles of length k can be written in any of the $c_k(\pi)!$ possible orders. So the number of ways of entering the numbers $1, \ldots, n$ giving rise to each permutation in the conjugacy class is $\prod k^{c_k(\pi)} c_k(\pi)!$.

The *cycle index* of the symmetric group S_n is the generating function for the numbers $c_k(\pi)$, for $k = 1, \ldots, n$. By convention it is normalised by dividing by $n!$. Thus,

$$Z(S_n) = \frac{1}{n!} \sum_{\pi \in S_n} \prod_{k=1}^{n} s_k^{c_k(\pi)}.$$

Because of the normalisation, this can be thought of as the probability generating function for the cycle structure of a random permutation: that is, the coefficient of the monomial $\prod s_k^{a_k}$ (where $\sum k c_k = n$) is the probability that a random permutation π has $c_k(\pi) = a_k$ for $k = 1, \ldots, n$ – this is

$$\frac{1}{\prod_k k^{a_k} a_k!}.$$

One result which we will meet later is the following. We adopt the convention that $Z(S_0) = 1$.

Proposition 3.22 $\sum_{n\geq 0} Z(S_n) = \exp\left(\sum_{k\geq 1} \frac{s_k}{k}\right)$.

Proof The left-hand side is equal to

$$\sum_{n\geq 0}\sum_{\sum a_k=n}\prod_{k\geq 1}\frac{s_k^{a_k}}{k^{a_k}a_k!} = \sum_{a_1,a_2,...}\prod_{k\geq 1}\frac{s_k^{a_k}}{k^{a_k}a_k!}$$

$$= \prod_{k\geq 1}\sum_{a\geq 0}\frac{s_k^a}{k^a a!}$$

$$= \prod_{k\geq 1}\exp\left(\frac{s_k}{k}\right)$$

$$= \exp\left(\sum_{k\geq 1}\frac{s_k}{k}\right)$$

as required. (The sum on the right-hand side of the first line is over all infinite sequences of natural numbers $(a_1, a_2, \ldots)$ with only finitely many entries non-zero.)

We will see much more about cycle indices in the chapter on orbit counting.

3.4 Lah numbers

The *Lah numbers*, sometimes called the 'Stirling numbers of the third kind', can be defined by

$$L(n,k) = \sum_{m=k}^{n}|s(n,m)|S(m,k).$$

Put $(x)^{(n)} = x(x+1)\cdots(x+n-1)$. Then we have

$$(x)^{(n)} = \sum_{m=1}^{n}|s(n,m)|x^m,$$

$$x^m = \sum_{k=1}^{m}S(m,k)(x)_k,$$

and so

$$(x)^{(n)} = \sum_{k=1}^{n}L(n,k)(x)_k,$$

by the rule for matrix multiplication. In other words, the Lah numbers express the 'rising factorials' $(x)^{(n)}$ in terms of the 'falling factorials' $(x)_n$.

Now

$$(x)^{(n+1)} = \sum L(n,k)(x)_k((x-k)+(n+k))$$
$$= \sum(n+k)L(n,k)(x)_k + \sum L(n,k)(x)_{k+1},$$

so

$$L(n+1,k) = (n+k)L(n,k) + L(n,k-1),$$

with the convention that $L(n,k) = 0$ if $k = 0$ or $k > n$. Now it is straightforward to show that

$$L(n,k) = \frac{n!}{k!}\binom{n-1}{k-1}$$

is the unique solution of this recurrence relation with the appropriate boundary conditions.

In other words, unlike the Stirling numbers, there is a closed form for the Lah numbers in terms of factorials.

3.5   More on formal power series

The enumeration of subsets and partitions makes an unexpected appearance in the rules for differentiating products and composites of formal power series. In fact, the formulae below work as well for n-times differentiable functions in the usual sense of calculus, since they depend only on the standard rules for differentiating sums and products and the Chain Rule.

For brevity, we use $f^{(n)}(x)$ for the result of differentiating $f(x)$ n times, and write $f'(x)$ for $f^{(1)}(x)$.

Products   The standard product rule

$$\frac{\mathrm{d}}{\mathrm{d}x}(f(x)g(x)) = \left(\frac{\mathrm{d}}{\mathrm{d}x}f(x)\right)g(x) + f(x)\left(\frac{\mathrm{d}}{\mathrm{d}x}g(x)\right)$$

extends to *Leibniz's rule*:

Proposition 3.23

$$\frac{\mathrm{d}^n}{\mathrm{d}x^n}(f(x)g(x)) = \sum_{k=0}^{n}\binom{n}{k}\left(\frac{\mathrm{d}^k}{\mathrm{d}x^k}f(x)\right)\left(\frac{\mathrm{d}^{n-k}}{\mathrm{d}x^{n-k}}g(x)\right).$$

Proof   The proof is by induction. The $n-1$st derivative is a sum of terms where f is differentiated k times and g $n-k-1$ times, with coefficient $\binom{n-1}{k}$. By the product rule, terms in $(\mathrm{D}^k f)(\mathrm{D}^{n-k}g)$ in $\mathrm{D}^n(fg)$ arise by differentiating terms in either $(\mathrm{D}^{k-1}f)(\mathrm{D}^{n-k}g)$ or $(\mathrm{D}^k f)(\mathrm{D}^{n-k-1}g)$, so the coefficient of this term is

$$\binom{n-1}{k-1} + \binom{n-1}{k} = \binom{n}{k}.$$

Taking $f(x) = e^{ax}$ and $g(x) = e^{bx}$, we obtain

$$(a+b)^n e^{(a+b)x} = \sum_{k=0}^{n} \binom{n}{k} (a^k e^{ax})(b^{n-k} e^{bx}),$$

so

$$(a+b)^n = \sum_{k=0}^{n} \binom{n}{k} a^k b^{n-k},$$

the Binomial Theorem for positive integer exponents. Similarly, taking $f(x) = x^a$ and $g(x) = x^b$, we obtain

$$(a+b)_{(n)} = \sum_{k=0}^{n} \binom{n}{k} a_{(k)} b_{(n-k)}.$$

Substitution   The Chain Rule tells us that

$$\frac{\mathrm{d}}{\mathrm{d}x}(f(g(x))) = f'(g(x))g'(x),$$

where, to attempt to minimise confusion, I have used $f'(x)$ for the derivative of $f(x)$ with respect to x, and $f'(g(x))$ for the result of substituting $g(x)$ for x in this derivative. As we have seen, the substitution of g into f is valid provided that $g(0) = 0$.

The generalisation of this to repeated derivatives is *Faà di Bruno's rule*.[2] If $a_1, \ldots, a_k$ are positive integers with sum n, let $P(n; a_1, \ldots, a_k)$ be the number of partitions of $\{1, \ldots, n\}$ into parts of size $a_1, \ldots, a_k$. Also, we use $f^{(k)}(x)$ for $\mathrm{D}^k(f(x))$. In the formula below, the summation is over all expressions for n as the sum of positive integers $a_1, \ldots, a_k$.

Proposition 3.24

$$\frac{\mathrm{d}^n}{\mathrm{d}x^n}(f(g(x))) = \sum_{a_1 + \cdots + a_k = n} P(n; a_1, \ldots, a_k) f^{(k)}(g(x)) g^{(a_1)}(x) \cdots g^{(a_k)}(x).$$

Proof   Again by induction. Suppose that we have a bijection between partitions of $\{1, \ldots, n\}$ and terms in the nth derivative of $f(g(x))$. When we differentiate the term $f^{(k)}(g(x)) g^{a_1}(x) \cdots g^{(a_k)}(x)$, corresponding to a partition of $\{1, \ldots, n\}$ into parts of sizes $a_1, \ldots, a_k$, we obtain $k+1$ terms:

- $f^{(k+1)}(g(x)) g^{a_1}(x) \cdots g^{(a_k)}(x) g'(x)$, corresponding to the partition of $\{1, \ldots, n+1\}$ in which $n+1$ is a singleton part;
- $f^{(k)}(g(x)) g^{a_1}(x) \cdots g^{(a_i+1)}(x) \cdots g^{(a_k)}(x)$, in which $n+1$ is adjoined to the ith part of the partition.

---

[2]Faà di Bruno is possibly the only mathematician who is also a Roman Catholic saint.

So each partition of $\{1,\ldots,n+1\}$ corresponds to a unique term in the sum, and we are done.

For example, we have

$$D^n(f(\exp(x)-1)) = \sum_{k=1}^{n} S(n,k)f^{(k)}(\exp(x)-1)\exp(kx),$$

since the sum of $P(n;a_1,\ldots,a_k)$ over all $(a_1,\ldots,a_k)$ with fixed n and k is just the number $S(n,k)$ of partitions with k parts. Putting $x=0$ we obtain the formula

$$b_n = \sum_{k=1}^{n} S(n,k)a_k$$

relating the coefficients of $f(x)$ and $f(\exp(x)-1)$.

3.6 Unimodality

For the Stirling numbers of the second kind the recurrence is

$$S(n,k) = S(n-1,k-1) + kS(n,k),$$

so to get any particular value we take the value immediately above, multiply it by its column number k, and add the value above and to the left:

1

1   1

1   3   1

1   7   6   1

$\downarrow 1\searrow\quad\downarrow 2\searrow\quad\downarrow 3\searrow\quad\downarrow 4\searrow$

1   15   25   10   1

For the unsigned Stirling numbers of the first kind the recurrence is

$$u(nk) = u(n-1,k-1) + (n-1)u(n-1,k),$$

so it works the same except that we multiply the value immediately above the one we want by its row number $n-1$ instead of its column number:

$$1$$

$$1 \qquad 1$$

$$2 \qquad 3 \qquad 1$$

$$6 \qquad 11 \qquad 6 \qquad 1$$
$$\downarrow 4 \searrow \quad \downarrow 4 \searrow \quad \downarrow 4 \searrow \quad \downarrow 4 \searrow$$
$$24 \qquad 50 \qquad 35 \qquad 10 \qquad 1$$

In both cases, we notice that the numbers in a row seem to increase to a maximum and then decrease. We now examine this property.

Given a sequence of positive numbers, say $a_0, a_1, a_2, \ldots, a_n$, we say that the sequence is *unimodal* if there is an index m with $0 \le m \le n$ such that

$$a_0 \le a_1 \le \cdots \le a_m \ge a_{m+1} \ge \cdots \ge a_n.$$

Example For fixed n, the binomial coefficients

$$\binom{n}{0}, \binom{n}{1}, \ldots, \binom{n}{n}$$

are unimodal. For we have

$$\binom{n}{k+1} = \frac{n-k}{k+1} \binom{n}{k},$$

so we have $\binom{n}{k+1}$ greater than, equal to or less than $\binom{n}{k}$ according as $n-k$ is greater than, equal to or less than $k+1$; that is, according as k is less than, equal to or greater than $(n+1)/2$. So, if $n = 2m$, we have

$$\binom{n}{0} < \binom{n}{1} < \cdots < \binom{n}{m} > \binom{n}{m+1} > \cdots > \binom{n}{n},$$

while if $n = 2m+1$, we have

$$\binom{n}{0} < \binom{n}{1} < \cdots < \binom{n}{m} = \binom{n}{m+1} > \binom{n}{m+2} > \cdots > \binom{n}{n}.$$

So the binomial coefficients increase to the middle and then decrease (remaining constant for one step if n is odd).

This is all very well, but it depends on having a formula for the binomial coefficients. We want to show that various other sequences are unimodal, so we need to develop some machinery. Here is a simple but useful test.

The sequence $a_0, a_1, a_2, \ldots, a_n$ of positive integers is said to be *log-concave* if $a_k^2 \geq a_{k-1}a_{k+1}$ for $1 \leq k \leq n-1$. The reason for the name is that the logarithms of the as are concave: setting $b_k = \log a_k$, we have $2b_k \leq b_{k-1} + b_{k+1}$, or in other words, $b_{k+1} - b_k \leq b_k - b_{k-1}$. So if we plot the points (k, b_k) for $0 \leq k \leq n$, then the slopes of the lines joining consecutive points decrease as k increases, so that the figure they form is concave when viewed from above.

Theorem 3.25 *A log-concave sequence is unimodal.*

Proof We have $a_k^2 \geq a_{k+1}a_{k-1}$; so $(a_{k+1}/a_k) \leq (a_k/a_{k-1})$. So if $a_k \leq a_{k-1}$ then $a_{k+1} \leq a_k$. So the numbers a_k increase until it first happens that $a_k \leq a_{k-1}$ and then decrease.

Theorem 3.26 *Let* $P(x) = \displaystyle\sum_{k=0}^{n} a_k x^k$ *be the generating polynomial for the numbers* $a_0, \ldots, a_n$. *Suppose that all the roots of the equation* $P(x) = 0$ *are real and negative. Then the sequence* $a_0, \ldots a_n$ *is log-concave.*

Proof The proof goes by induction on n. If $n = 1$, then there are only two numbers, and trivially they form a log-concave sequence. We will look at the case $n = 2$ before going on to the general case.

The coefficients of the polynomial $P(x)$ are, by assumption, all positive, so there cannot be a non-negative root. So our assumption is that the roots are all real. Now for $n = 2$ we have $P(x) = a_0 + a_1 x + a_2 x^2$; this has real roots if and only if its discriminant is non-negative, that is, if and only if $a_1^2 - 4a_0 a_2 \geq 0$; but this implies that $a_1^2 \geq a_2 a_0$, which is the definition of log-concavity!

Now we turn to the general case. Suppose that $P(x) = (x + c)Q(x)$, where $c > 0$ and

$$Q(x) = b_{n-1}x^{n-1} + \cdots + b_1 x + b_0.$$

Now the polynomial $Q(x)$ has all its roots real and negative, since they are all the roots of $P(x)$ except for $-c$. By the inductive hypothesis, the sequence $b_0, \ldots, b_{n-1}$ is log-concave; that is,

$$b_k^2 \geq b_{k-1}b_{k+1}$$

for $k = 1, \ldots, n-2$. Also, since $P(x) = (x + c)Q(x)$, we have $a_0 = cb_0$, $a_n = b_{n-1}$, and $a_k = b_{k-1} + cb_k$ for $1 \leq k \leq n-1$.

We first show that $b_k b_{k-1} \geq b_{k+1}b_{k-2}$ for $2 \leq k \leq n-2$. For we have

$$b_k^2 b_{k-1} \geq b_{k+1}b_{k-1}^2 \geq b_{k+1}b_k b_{k-2};$$

dividing by b_k gives the result.

Now for $2 \leq k \leq n-2$, we have

$$
\begin{aligned}
a_k^2 - a_{k+1}a_{k-1} &= (b_{k-1} + cb_k)^2 - (b_k + cb_{k+1})(b_{k-2} + cb_{k-1}) \\
&= (b_{k-1}^2 - b_k b_{k-2}) + c(b_{k-1}b_k - b_{k+1}b_{k-2}) + c^2(b_k^2 - b_{k+1}b_{k-1});
\end{aligned}
$$

and all three terms are non-negative since $c > 0$.

We have to check also the cases $k = 1$ and $k = n-1$. For $k = 1$ we have $a_1^2 - a_2 a_0 = (b_0 + cb_1)^2 - (b_1 + cb_2)cb_0$; for $k = n-1$, we have $a_{n-1}^2 - a_n a_{n-2} = (b_{n-2} + cb_{n-1})^2 - b_{n-1}(b_{n-3} + cb_{n-2})$. In each case, it is easy to check that the expression is non-negative.

So the induction step is proved.

Example 1: Binomial coefficients   By the Binomial Theorem, we have

$$
\sum_{k=0}^{n} \binom{n}{k} x^k = (1+x)^n,
$$

and the equation $(1+x)^n = 0$ has just the root -1 with multiplicity n. So the binomial coefficients are log-concave and hence unimodal.

Example 2: Stirling numbers of the first kind   Let $u(n,k)$ be the number of permutations of $\{1, \ldots, n\}$ having k cycles. We showed earlier that

$$
\sum_{k=1}^{n} u(n,k) x^k = x(x+1) \cdots (x+n-1),
$$

and the polynomial on the right has roots $0, -1, -2, \ldots, -(n-1)$. We can neglect the zero root: the Stirling numbers start at $k = 1$ rather than zero, and dividing by x simply changes the indexing so that they start at 0. So again the Stirling numbers are log-concave and hence unimodal.

Example 3: Stirling numbers of the second kind   Let $S(n,k)$ be the number of partitions of $\{1, \ldots, n\}$ into k parts. These numbers are also unimodal. The proof is a little more difficult, but is a good showcase for some elementary real analysis (Rolle's Theorem).

We begin with the recurrence relation for the Stirling numbers of the first kind, proved in Proposition 3.7:

$$
S(n,1) = S(n,n) = 1, \qquad S(n,k) = S(n-1,k-1) + kS(n-1,k) \text{ for } 1 < k < n.
$$

Let

$$
P_n(x) = \sum_{k=0}^{n} S(n,k) x^k.
$$

We have $P_0(x) = 1$. For $n > 0$, we have $S(n,0) = 0$, so zero is a root of $P_n(x) = 0$. We have to show that the other roots are real and negative; then Theorem 3.26 above applies to $P_n(x)/x$. We prove this by induction: $P_1(x) = x$ has a single root at $x = 0$, while $P_2(x) = x + x^2$ has roots at $x = 0$ and $x = -1$; so the induction begins.

From the recurrence relation, we have

$$\begin{aligned}
P_n(x) &= \sum_{k=1}^{n} S(n,k)x^k \\
&= \sum_{k=1}^{n} S(n-1,k-1)x^k + \sum_{k=1}^{n} kS(n-1,k)x^k \\
&= x\left(dP_{n-1}(x)/dx + P_{n-1}(x)\right).
\end{aligned}$$

Putting $Q_n(x) = P_n(x)e^x$, we see that $P_n(x) = 0$ and $Q_n(x) = 0$ have the same roots. The identity above, multiplied by e^x, gives

$$x\,dQ_{n-1}(x)/dx = Q_n(x).$$

By Rolle's Theorem, there is a root of $Q_n(x)$ between each pair of roots of $Q_{n-1}(x)$, and one to the left of the smallest root of $Q_{n-1}(x)$ (since $Q_{n-1}(x) \to 0$ as $x \to -\infty$); and also a a root at 0. This accounts for $(n-2) + 1 + 1$ roots, that is, all the roots of $Q_n(x)$. So the induction step is complete.

3.7   Appendix: Exponential and logarithm

The exponential and logarithmic series are inverses of each other, under the operation of substitution: that is,

$$\exp(\log(1+x)) = 1+x.$$

We give here a combinatorial proof, rather than the usual analytic proof. We will see that the proof ultimately rests on facts about the sign of a permutation.

First reduction   We are trying to show that $\exp(\log(1+x)) = 1+x$ as formal power series, that is,

$$1 + \left(x - \frac{x^2}{2} + \cdots\right) + \frac{\left(x - \frac{x^2}{2} + \cdots\right)^2}{2!} + \cdots = 1+x.$$

It is easy to work out the first few terms explicitly and check them. To prove the result, we have to show that the coefficient of x^n on the left-hand side is zero for $n > 1$. We show first:

The coefficient of x^n in $\exp(\log(1+x))$ is

$$\sum_{k=1}^{n}(-1)^{n-k}T(n,k)/k!,$$

where $T(n,k)$ is calculated as follows: for each expression $n = a_1 + \cdots + a_k$ as an ordered sum (a *composition*) of k positive integers, take the product of the reciprocals of $a_1,\ldots,a_k$; then add these products.

For example, if $n = 4$ and $k = 2$, there are three expressions, namely $1+3$, $2+2$, $3+1$; so

$$T(4,2) = \frac{1}{3} + \frac{1}{4} + \frac{1}{3} = \frac{11}{12}.$$

To see this note that the coefficient of x^n in $\exp(\log(1+x))$ is the sum (over k) of its coefficient in $(x - x^2/2 + \cdots)^k$, divided by $k!$. The latter has a coefficient, for each expression $n = a_1 + \cdots + a_k$, obtained by taking the coefficient of x^{a_i} in the ith factor. This coefficient is $(-1)^{a_i-1}/a_i$, and the product of these is $(-1)^{n-k}$ times the product of the reciprocals of the a_i, as required.

Expression in terms of permutations To show the required identity, we show that

$$T(n,k) = \frac{k!\,u(n,k)}{n!}.$$

For example, there are 11 permutations of $\{1,\ldots,4\}$ with two cycles; eight of these consist of a fixed point and a 3-cycle, and three are products of two 2-cycles. Then $11 \cdot 2!/4! = T(4,2)$.

We actually prove a 'local' version. Suppose that $b_1,\ldots,b_n$ are non-negative integers with $b_1 + 2b_2 + \cdots + nb_n = n$ and $b_1 + b_2 + \cdots + b_n = k$. We show that the contribution to $T(n,k)$ from expressions for n as a sum of b_1 ones, b_2 twos, $\ldots$ is equal to $k!/n!$ times the number of permutations with b_i cycles of length i for $i = 1,\ldots,n$, which was computed earlier in Proposition 3.21.

For the contribution of such sums to $T(n,k)$ is equal to the number of ways of ordering the summands times the contribution from each summand. The number of orderings is

$$\frac{k!}{b_1!b_2!\cdots b_n!},$$

the denominator accounting to the fact that reordering identical summands does not change the expression. Of course the value of each summand is

$$\frac{1}{1^{b_1}2^{b_2}\cdots n^{b_n}}.$$

To count permutations, we write out the 'pattern' consisting of b_i cycles of length i, for $i = 1,\ldots,n$. We can write the numbers $1,\ldots,n$ into the pattern in

$n!$ different ways. But each permutation arises in several different ways, for two reasons: we may start each cycle at any point (giving a factor $1^{b_1}2^{b_2}\cdots n^{b_n}$); and we may rearrange the cycles of the same length (giving a factor $b_1!\,b_2!\cdots b_n!$). So the number of such permutations is

$$\frac{n!}{1^{b_1}b_1!\,2^{b_2}b_2!\cdots n^{b_n}b_n!}.$$

By inspection, we see that the claim is proved.

Completion of the proof   We have to evaluate

$$\sum_{k=1}^{n}(-1)^{n-k}T(n,k)/k!\,.$$

By the result of the last section, this is equal to

$$\sum_{k=1}^{n}(-1)^{n-k}u(n,k)/n! = \frac{1}{n!}\sum_{\pi\in S_n}\mathrm{sgn}(\pi)=0$$

for $n>1$, since there are equally many permutations with each sign.

3.8   Exercises

3.1 Write down Pascal's triangle mod 2; that is, record only whether each entry is odd or even. You should find that the triangle has a fractal structure; can you explain why? [Exercise 3.6 below may help.]

3.2 (A few binomial coefficient identities.) Prove the following:

(a) $\displaystyle\sum_{k=0}^{n}(-1)^k\binom{n}{k}=0.$

(b) $\displaystyle\sum_{k=0}^{n}\binom{n}{k}^2=\binom{2n}{n}.$

(c) $\displaystyle\sum_{k=0}^{n}(-1)^k\binom{n}{k}^2=\begin{cases}0 & \text{if } n \text{ is odd,}\\ (-1)^m\binom{2m}{m} & \text{if } n=2m.\end{cases}$

3.3 Prove that the generating function for the central binomial coefficients is

$$\sum_{n\ge 0}\binom{2n}{n}x^n=(1-4x)^{-1/2},$$

and deduce that

$$\sum_{k=0}^{n}\binom{2k}{k}\binom{2(n-k)}{n-k}=4^n.$$

Remark A bijective proof of the last equality is quite challenging!

3.4 This exercise gives a technique which can in principle be generalised to sum the binomial coefficients $\binom{n}{k}$ with k running over any arithmetic progression.

(a) Show that

$$\sum_{k=0}^{\lfloor n/2 \rfloor} \binom{n}{2k} = \sum_{k=0}^{\lfloor (n-1)/2 \rfloor} \binom{n}{2k+1} = 2^{n-1}.$$

(b) By applying the Binomial Theorem to $(1+\mathrm{i})^n$, where i is a square root of -1, and using the fact that $1+\mathrm{i} = \sqrt{2}\mathrm{e}^{\mathrm{i}\pi/4}$, find a formula for

$$\sum_{k=0}^{\lfloor n/4 \rfloor} \binom{n}{4k}.$$

(You should distinguish cases depending on the congruence of n modulo 8.)

3.5 Let p be a prime number.

(a) Show that $\binom{p}{k}$ is congruent to zero (mod p) for $1 \leq k \leq p-1$.

(b) Hence show that $(x+y)^p \equiv x^p + y^p$ (mod p).

(c) Deduce *Fermat's Little Theorem*: $n^p \equiv n$ (mod p) for any natural number n.

3.6 Again let p be a prime number. Using the result of the previous exercise, show the following. Suppose that $n = ap+b$ and $k = cp+d$, with $0 \leq b, d \leq p-1$. By computing the coefficient of x^{cp+d} in $(1+x)^{ap+b}$, show that

$$\binom{ap+b}{cp+d} \equiv \binom{a}{b}\binom{c}{d} \text{ (mod } p\text{)}.$$

Deduce *Lucas' Theorem*: If $n = a_r p^r + a_{r-1} p^{r-1} + \cdots + a_0$ and $k = b_r p^r + b_{r-1} p^{r-1} + \cdots + b_0$, where $0 \leq a_i, b_i \leq p-1$ for $i = 0, \ldots, r$, then

$$\binom{n}{k} \equiv \binom{a_r}{b_r}\binom{a_{r-1}}{b_{r-1}} \cdots \binom{a_0}{b_0} \text{ (mod } p\text{)}.$$

In particular, show that $\binom{n}{k}$ is divisible by p if and only if $b_i > a_i$ for some value of i.

3.7 Deduce the Binomial Theorem by applying Taylor's Theorem to the analytic function $(1+x)^a$ (for arbitrary real number a), using the standard formula for the derivative of $(1+x)^a$.

3.8 Prove that, if n is even, then

$$\frac{2^n}{n+1} \leq \binom{n}{n/2} \leq 2^n.$$

Use Stirling's formula to prove that

$$\binom{n}{n/2} \sim \frac{2^n}{\sqrt{\pi n/2}}.$$

How accurate is this estimate for small n?

3.9 Use the method of the preceding exercise, together with the Central Limit Theorem, to deduce the constant in Stirling's formula.

3.10 Prove directly that, if $0 \leq k < n$, then

$$\sum_m (-1)^{m-k} \binom{n}{m}\binom{m}{k} = \sum_m (-1)^{n-m}\binom{n}{m}\binom{m}{k} = 0.$$

3.11 Formulate and prove an analogue of Proposition 3.12 for binomial coefficients.

3.12 In how many ways can k identical sweets be distributed to n children if the number x_i of sweets given to the ith child is required to satisfy $x_i \geq a_i$, for some numbers $a_1, \ldots, a_n$?

3.13 Let B be the infinite matrix whose (n, k) entry is $\binom{n}{k}$. What are the entries of the matrix B^2?

3.14 Let $k_1, k_2, \ldots, k_r$ be non-negative integers with sum n. The *multinomial coefficient* is defined as

$$\binom{n}{k_1, k_2, \ldots, k_r} = \frac{n!}{k_1! k_2! \ldots k_r!}.$$

Prove that

(a) The number of ways of colouring elements of an n-set with colours $c_1, \ldots, c_r$ so that there are k_i objects of colour c_i is $\binom{n}{k_1, k_2, \ldots, k_r}$;

(b) If the elements are so coloured, and the elements of each colour are ordered, the number of orderings of the whole set inducing the given ordering of the elements of each colour is $\binom{n}{k_1, k_2, \ldots, k_r}$.

Can you find a bijective proof that the numbers in the two parts are the same?

3.15 Prove the *Multinomial Theorem*:

$$(x_1 + x_2 + \cdots + x_r)^n = \sum_{k_1+k_2+\cdots+k_r=n} \binom{n}{k_1, k_2, \ldots, k_r} x_1^{k_1} x_2^{k_2} \cdots x_r^{k_r}.$$

3.16 Let $B(n)$ be the number of partitions of $\{1, \ldots, n\}$. Prove that

$$\sqrt{n!} \leq B(n) \leq n!.$$

3.17 Prove that $\log n!$ is greater than $n \log n - n + 1$ and differs from it by at most $\frac{1}{2} \log n$. Deduce that

$$\frac{n^n}{e^{n-1}} \leq n! \leq \frac{n^{n+1/2}}{e^{n-1}}.$$

3.18 Prove that

$$(-1)^{n-k} \binom{n}{k} = \binom{-n+k-1}{k}$$

for $0 \leq k \leq n$. Use this and Proposition 3.3 to prove the Binomial Theorem for negative integer exponents.

3.19   (a) Prove Proposition 3.16.

  (b) Prove Proposition 3.15.

3.20 Let $A(x_1, \ldots, x_n)$ be the polynomial in indeterminates $x_1, \ldots, x_n$ given by

$$A(x_1, \ldots, x_n) = \prod_{1 \leq i < j \leq n} (x_j - x_i).$$

Prove that, if two of the indeterminates are transposed and the others are fixed, then $A(x_1, \ldots, x_n)$ is mapped to its negative.

Hence show that, if we define a function s from the symmetric group to the group of integers mod 2 by the rule that

$$A(x_{1\pi}, x_{2\pi}, \ldots, x_{n\pi}) = (-1)^{s(\pi)} A(x_1, \ldots, x_n),$$

then s is a group homomorphism, and $s(\pi) = -1$ if π is a transposition.

Hence show that, if $n \geq 2$, then the set of elements $\pi \in S_n$ for which $s(\pi) = -1$ is a normal subgroup of index 2 (that is, exactly half of the elements of S_n have $s(\pi) = -1$). This normal subgroup is the *alternating group* A_n.

3.21 Prove that

$$\sum_{n \geq k} \frac{s(n,k)x^n}{n!} = \frac{(\log(1+x))^k}{k!}$$

for $k \geq 1$. What happens when this equation is summed over k?

3.22 What is the relation between the numbers $T(n,k)$ defined in Section 3.7 and Stirling numbers?

3.23 A *total preorder* is a reflexive and transitive relation on a set X which satisfies the *trichotomy law*, that is, for any $x,y \in X$, either $R(x,y)$ or $R(y,x)$ holds.

  (a) Show that the number of total preorders of an n-set is

$$P(n) = \sum_{k=1}^{n} S(n,k)k! \, .$$

  (b) Deduce that the exponential generating function for $(P(n))$ is $1/(2-\exp(x))$.

3.24 Prove that the number $P(n)$ of total preorders on n points is given by the formula

$$P(n) = \sum_{k \geq 1} \frac{k^n}{2^{k+1}}.$$

[Hint: $k^n = \sum_{i=1}^{n} S(n,i)(k)_i$.]

  Show that the following method allows us to choose a random preorder on n points uniformly:

  (a) choose a positive integer k from the probability distribution K given by

$$\mathbb{P}(K = k) = \frac{k^n}{P(n)2^{k+1}};$$

  (b) assign a random score $s(x) \in \{1, \ldots, k\}$ to each point x;
  (c) put $x \leq y$ if and only if $s(x) \leq s(y)$.

(This method is due to H. Maassen and T. Bezembinder.)

3.25 (a) Prove the formula for the Lah numbers given in the text.
  (b) Show that $L(n,k)$ is the number of ways of partitioning the set $\{1, \ldots, n\}$ into k non-empty parts and imposing a linear order on each part. (Contrast this with the preorder numbers $P(n)$ which count the number of partitions with a linear order on the set of parts.)

3.26 Prove that the generating function for the central binomial coefficients is

$$\sum_{n \geq 0} \binom{2n}{n} x^n = (1 - 4x)^{-1/2},$$

and deduce that

$$\sum_{k=1}^{n} \binom{2k}{k} \binom{2(n-k)}{n-k} = 4^n.$$

[Note: Finding a counting proof of this identity is quite challenging!]

3.27 Find a formula for the number $P(n; a_1, \ldots, a_k)$ appearing in Faà di Bruno's formula.

3.28   (a) Let p_n be the counting function for permutations on an even number of points: that is, $p_n = n!$ if n is even, $p_n = 0$ if n is odd. Show that the exponential generating function of the sequence (p_n) is $P(x) = (1 - x^2)^{-1}$.

(b) Let $n = 2m$. Show that the number of permutations of $\{1, \ldots, n\}$ with all cycles even is the square of the number of permutations with all cycles of length 2. (Hint: if σ and τ are permutations with all cycles of length 2, colour their cycles red and blue respectively, and produce a permutation with all cycles even by following red and blue edges alternately. Show that every pair (σ, τ) gives rise to 2^c permutations, where c is the number of cycles of length greater than 2, while every permutation with all cycles even and c of length greater than 2 arises from 2^c such pairs (σ, τ).)

(c) Deduce that, if e_n is the number of permutations of $\{1, \ldots, n\}$ with all cycles even, then the exponential generating function for (e_n) is $E(x) = (1 - x^2)^{-1/2}$.

(d) Let o_n be the number of permutations of a set $\{1, \ldots, n\}$ with all cycles odd, where n is even (that is, $o_n = 0$ if n is odd). Show that the exponential generating function $O(x)$ for the sequence (o_n) satisfies $E(x)O(x) = P(x)$, by decomposing a permutation into the product of a permutation with all cycles even and one with all cycles odd.

(e) Deduce that $e_n = o_n$ for all even numbers n.
This demonstrates a result mentioned in the Preface.

Remark   Can you find a bijective proof of the last assertion? [See the paper by Richard Lewis and Simon Norton for a crib.]

3.29 Give a treatment of the identity

$$\log(1 + (\exp(x) - 1)) = x$$

similar to that for $\exp(\log(1 + x)) = 1 + x$ in the appendix to this chapter.

3.8.1   A prize question

In the course, a prize was offered for the first solution to this question.

The following problem arises in the theory of clinical trials. A new drug is to be tested out. Of $2n$ subjects in the trial, n will receive the new drug and n will get a placebo. To avoid bias, it is important that the doctor administering the treatments does not know, and cannot reliably guess, which treatment each patient receives. The patients enter the trial one at a time, and are numbered from 1 to $2n$.

If the treatments were allocated randomly with probability $1/2$, the doctor's guesses could be no better than random (so that the expected values for the numbers of correct and incorrect guesses are both n); but then the numbers of patients receiving drug and placebo would be unlikely to be equal. Given that they must balance, the doctor can certainly guess at least the last patient's treatment correctly.

If we allocated the drug and the placebo randomly to patients $2i - 1$ and $2i$ for $i = 1, \ldots, n$, then the doctor can correctly guess the treatment for each even-numbered patient.

Suppose that instead we choose a random set of n patients to allocate the drug to, and the remaining n get the placebo; each of the $\binom{2n}{n}$ sets is equally likely. Suppose also that the doctor guesses according to the following rule. If the number of patients so far having the drug and the placebo are equal, he guesses at random about the next treatment. If the drug has occurred more often than the placebo, he guesses that the next treatment is the placebo, and *vice versa* if the placebo has occurred more often than the drug.

Find a formula, and an asymptotic estimate, for the expected value of the difference between the number of correct guesses and the number of incorrect guesses that the doctor makes.

Solution   We use the result of Problem 3.3 above, the identity

$$\sum_{k=0}^{n} \binom{2k}{k} \binom{2(n-k)}{n-k} = 2^{2n}.$$

Following the hint, we first calculate the expected number of times during the trial when the numbers of patients receiving drug and placebo are equal. This is obtained by summing, over all n-element subsets A of $\{1, \ldots, 2n\}$, the number of values of k for which $|A \cap \{1, \ldots, 2k\}| = k$, and dividing by the number $\binom{2n}{n}$ of subsets. Now the sum can be calculated by counting, for each value of k, the number of n-subsets A for which $|A \cap \{1, \ldots, 2k\}| = k$, and summing the result over k.

For a given k, the number of subsets is $\binom{2k}{k} \binom{2(n-k)}{n-k}$, since we must choose k of the numbers $1, \ldots, 2k$, and $n - k$ of the numbers $2k + 1, \ldots, 2n$. Hence, by the stated result, the sum is 2^{2n}, and the average is $2^{2n}/\binom{2n}{n}$.

Now consider the doctor's guesses in any particular trial. At any stage where equally many patients have received drug and placebo, he guesses at random, and is equally likely to be right as wrong. Such points contribute zero to the expected number of correct minus incorrect guesses. In each interval between two consecutive such stages, say $2k$, and $2l$, the doctor will guess right one more time than he guesses wrong. (For example, if the $2k$th patient gets the drug, then between stages $2k+1$ and $2l$ the number of patients getting the drug is $l-k-1$ and the number getting the placebo is $l-k$, but the doctor will always guess the placebo.) So the expected number of correct minus incorrect guesses is the number of such intervals, which is one less than the number of times that the numbers are equal.

So the expected number is $2^{2n}/\binom{2n}{n} - 1$, which is asymptotically $\sqrt{\pi n}$, by the result of Exercise 3.8.

CHAPTER 4

Recurrence relations

A recurrence relation expresses the nth term of a sequence as a function of the preceding terms. The most general form of a recurrence relation takes the form

$$u_n = F_n(u_0, \ldots, u_{n-1}) \text{ for } n \geq 0.$$

Clearly such a recurrence has a unique solution. (Note that this allows the possibility of prescribing some initial values, by choosing the first few functions to be constant. Also, it is permissible for the function F_n to depend explicitly on n.)

In this chapter, we will describe the complete solution of linear recurrence relations of finite length with constant coefficients. For other types, we usually do not give any general theory, but describe some important examples and methods for solution. Two of these examples, the *partition numbers* and the *Catalan numbers*, are of such importance that we devote a section to further study of each.

To begin, an example to show that the first recurrence relation we think of for a sequence of numbers is not always the easiest!

Example: Compositions of an integer   In how many ways is it possible to write the positive integer n as a sum of positive integers, where the order of the summands is significant? (In the case where the order is not significant, we have a *partition* of n; to distinguish the case where order is significant, we call such an expression a *composition* of n.

For example, there are five partitions of 4, namely

$$4 = 3+1 = 2+2 = 2+1+1 = 1+1+1+1.$$

Of these, $3+1$ can be re-ordered as $1+3$, and $2+1+1$ as either $1+2+1$ or $1+1+2$. So there are eight compositions of 4.

Let u_n be the number of compositions of n. One possible composition has a single summand n. In any other expression, if $n - i$ is the first summand, then it is followed by an expression for i as an ordered sum, of which there are u_i possibilities. Thus

$$u_n = 1 + u_1 + u_2 + \cdots + u_{n-1},$$

for $n \geq 1$. (When $n = 1$, this reduces to $u_1 = 1$.) This is a recurrence relation which determines u_n for all $n \geq 1$.

Since

$$u_{n-1} = 1 + u_1 + u_2 + \cdots + u_{u-2},$$

the recurrence reduces to the much simpler form

$$u_n = 2u_{n-1} \text{ for } n > 1,$$

with initial condition $u_1 = 1$. This obviously has the solution $u_n = 2^{n-1}$ for $n \geq 1$.

The simpler recurrence can be proved directly. Starting from a composition of $n - 1$, we obtain two compositions of n: one by adding a summand 1 at the beginning, and the other by increasing the first summand by 1. There is no overlap, since the first construction gives all compositions beginning with 1, and the second gives all beginning with a number greater than 1; moreover, every composition of n is obtained from one or other construction.

4.1 Linear recurrences with constant coefficients

A sequence (u_n) satisfies a linear recurrence with constant coefficients if there is another sequence (a_n) (for $n \geq 1$) such that

$$u_n = \sum_{k=1}^{n} a_n u_{n-k}$$

for all $n \geq n_0$.

This can be expressed in terms of generating functions as follows. Let $U(x) = \sum_{n \geq 0} u_n x^n$ and $A(x) = 1 - \sum_{n \geq 1} a_n x^n$.

Proposition 4.1 *The sequence* (u_n) *satisfies*

$$u_n = \sum_{k=1}^{n} a_n u_{n-k}$$

for all $n \geq n_0$ *if and only if the generating functions defined above satisfy the condition that* $U(x)A(x)$ *is a polynomial of degree less than* n_0.

Proof The recurrence expresses the condition that the coefficient of x^n in $U(x)A(x)$ is zero for $n \geq n_0$.

4.1.1   Bounded recurrences

One type of linear recurrence which can be solved completely is of the form

$$u_n = a_1 u_{n-1} + a_2 u_{n-2} + \cdots + a_k u_{n-k} \tag{4.1}$$

for $n \geq k$, where the k values $u_0, u_1, \ldots, u_{k-1}$ are prescribed.

If we consider the recurrence (4.1) without the initial values, we see that sums and scalar multiples of solutions are solutions. So, taking sequences over a field such as the rational numbers, we see that the set of solutions is a vector space over the field. Its dimension is k, since the k initial values can be prescribed abitrarily.

Thus, if we can write down k linearly independent solutions, the general solution is a linear combination of them.

We can find very easily the generating function for a solution of this recurrence, as follows.

The *characteristic polynomial* of the recurrence (4.1) is defined to be the polynomial

$$A(x) = 1 - a_1 x - a_2 x^2 - \cdots - a_k x^k.$$

Now we have the following result:

Theorem 4.2 *Let* $(u_0, u_1, \ldots)$ *be a sequence of numbers, with generating function* $\sum u_n x^n = U(x)$, *and let* n_0 *be an integer satisfying* $n_0 \geq k$. *Then the following conditions are equivalent:*

(a) *the sequence* (u_n) *satisfies the recurrence relation (4.1) for* $n \geq n_0$;

(b) $A(x)U(x)$ *is a polynomial of degree less than* n_0, *where* $A(x)$ *is the characteristic polynomial of (4.1).*

Proof As in Proposition 4.1, the proof is immediate from the observation that the relation (4.1) asserts that the coefficient of x^n in $A(x)U(x)$ is zero.

Corollary 4.3 *If a sequence* (u_n) *satisfies the recurrence (4.1) for all* $n \geq k$, *then its generating function is given by* $U(x) = P(x)/A(x)$, *where* $A(x)$ *is the characteristic polynomial of the recurrence, and* $P(x)$ *is an arbitrary polynomial of degree less than* k.

The k coefficients in $P(x)$ are the independent parameters in the solution of the recurrence, and can be found using the initial values $u_0, \ldots, u_{k-1}$ if these are prescribed.

Now we want to use the generating function to find an explicit form for the coefficients.

Theorem 4.4 *Suppose that the recurrence relation (4.1) has characteristic polynomial*

$$A(x) = (1 - \alpha_1 x)^{m_1} \cdots (1 - \alpha_r x)^{m_r},$$

where $\alpha_1, \ldots, \alpha_r$ are distinct complex numbers and $m_1, \ldots, m_r$ are positive integers. (Thus, we have $m_1 + \cdots + m_r = k$.) Suppose that the sequence (u_n) satisfies the relation (4.1) for all $n \geq k$. Then

$$u_n = p_1(n)\alpha_1^n + \cdots + p_r(n)\alpha_r^n,$$

where p_i is a polynomial of degree at most $m_i - 1$.

Remark The roots $\alpha_1, \ldots, \alpha_r$ are the inverses of the roots of the characteristic polynomial of the recurrence; so they are the roots of the 'inverse polynomial'

$$x^k - a_1 x^{k-1} - a_2 x^{k-2} - \cdots - a_k = 0.$$

Proof We have

$$\sum u_n x^n = U(x) = \frac{P(x)}{A(x)}.$$

We use the technique of partial fractions to evaluate this fraction.

First, if the degree of P is less than k, we claim that

$$\frac{P(x)}{A(x)} = \frac{P_1(x)}{(1 - \alpha_1 x)^{m_1}} + \cdots + \frac{P_r(x)}{(1 - \alpha_r x)^{m_r}},$$

where the degree of P_i is less than m_i for $i = 1, \ldots, r$.

To prove this, let

$$R_i(x) = \prod_{j \neq i}(1 - \alpha_j x)^{m_j} = A(x)/(1 - \alpha_i x)^{m_i}$$

for $i = 1, \ldots, r$. The polynomials $R_1(x), \ldots, R_r(x)$ have greatest common divisor 1, so by Euclid's algorithm there exist polynomials $S_1(x), \ldots, S_r(x)$ such that

$$S_1(x)R_1(x) + \cdots + S_r(x)R_r(x) = 1.$$

Multiplying by $P(x)$, and putting $T_i(x) = S_i(x)P(x)$, we have

$$T_1(x)R_1(x) + \cdots + T_r(x)R_r(x) = P(x).$$

Now let $P_i(x)$ be the remainder when $T_i(x)$ is divided by $(1 - \alpha_i x)^{m_i}$. Then

$$P_1(x)R_1(x) + \cdots + P_r(x)R_r(x) \equiv P(x) \bmod A(x)$$

and since both left and right sides of this congruence have degree strictly less than $\sum m_i = k$, we have equality. Now dividing by $A(x)$ gives the claimed result.

Next we show that, if $P(x)$ is a polynomial of degree less than m, then

$$\frac{P(x)}{(1-\alpha x)^m} = \sum_{j=0}^{m-1} \frac{c_j}{(1-\alpha x)^j}$$

for some uniquely determined numbers $c_0, \ldots, c_{m-1}$. To see this, multiply both sides by $(1-\alpha x)^m$ to obtain an equation between two polynomials, the right-hand side involving the coefficients $c_0, \ldots, c_{m-1}$; equating coefficients of powers of x from largest to smallest now determines these coefficients.

To conclude the proof, we observe that, by the Binomial Theorem with negative exponent, we have

$$\begin{aligned}
(1-\alpha x)^{-j} &= \sum_{n \geq 0} \binom{-j}{n}(-\alpha x)^n \\
&= \sum_{n \geq 0} \binom{j+n-1}{n}(\alpha x)^n \\
&= \sum_{n \geq 0} \binom{j+n-1}{j-1}\alpha^n x^n,
\end{aligned}$$

and the binomial coefficient $\binom{j+n-1}{j-1}$ is a polynomial of degree $j-1$ in n. Taking a linear combination of such expressions for $j = 1, \ldots, m$ gives a coefficient $P(n)\alpha^n$, where P is a polynomial of degree at most $m-1$.

An alternative proof is outlined in Exercise 4.1.

In particular, we see that for a recurrence of depth 2, that is, $k = 2$, there are just two possibilities:

(a) the characteristic polynomial is $(1-\alpha x)(1-\beta x)$ with $\alpha \neq \beta$, and the solution is $u_n = A\alpha^n + B\beta^n$;

(b) the characteristic polynomial is $(1-\alpha x)^2$, and the solution is $(An+B)\alpha^n$.

Example Consider the recurrence $u_n = 2u_{n-1} - u_{n-2}$ for $n \geq 2$.

The characteristic polynomial is $1 - 2x + x^2 = (1-x)^2$; so the general solution is $u_n = An + B$. The coefficients A and B could be determined from the values of u_0 and u_1 if these were given.

Example: Fibonacci numbers Consider the Fibonacci recurrence

$$F_n = F_{n-1} + F_{n-2} \text{ for } n \geq 2.$$

The characteristic polynomial is

$$1 - x - x^2 = (1 - \alpha x)(1 - \beta x)$$

where $\alpha, \beta = (1 \pm \sqrt{5})/2$. So the general solution is

$$F_n = A\alpha^n + B\beta^n,$$

and A and B can be determined from the initial conditions.

If we take as initial conditions $F_0 = F_1 = 1$, we obtain the two equations

$$
\begin{aligned}
A + B &= 1, \\
A\alpha + B\beta &= 1.
\end{aligned}
$$

Solving these equations gives the solution we found earlier.

4.1.2 Sequences with forbidden subwords

This section describes a method due to Guibas and Odlyzko for counting sequences of zeros and ones not containing a prescribed finite sequence as a consecutive subsequence.

Let a be a binary sequence of length k. How many binary sequences of length n do not contain a as a consecutive subword?

Suppose, for example, that $a = 11$, so that we are counting binary strings with no two consecutive ones. Let $f(n)$ denote the number of such sequences of length n, and let $g(n)$ the number of sequences commencing with 11 but having no other occurrence of 11. Then

$$2f(n) = f(n+1) + g(n+1),$$

since if we take a string with no occurrence of 11 and precede it with a 1, then the only possible position of 11 is at the beginning. Also, if we take a string with no occurrence of 11 and precede it with 11, then the resulting sequence contains 11, but possibly two occurrences (if the original string began with a 1); so we have

$$f(n) = g(n+1) + g(n+2).$$

Then $f(n) = (2f(n) - f(n+1)) + (2f(n+1) - f(n+2))$, so we have the Fibonacci recurrence

$$f(n+2) = f(n) + f(n+1).$$

Since $f(0) = 1 = F_1$ and $f(1) = 2 = F_2$, a simple induction proves that $f(n) = F_{n+1}$ for all $n \geq 0$.

Guibas and Odlyzko extended this approach to arbitrary forbidden substrings. They defined the *correlation polynomial* of a binary string a of length k to be

$$C_a(x) = \sum_{j=0}^{k-1} c_a(j) x^j,$$

where $c_a(0) = 1$ and, for $1 \leq j \leq k-1$,

$$c_a(j) = \begin{cases} 1 & \text{if } a_1 a_2 \cdots a_{k-j} = a_{j+1} a_{j+2} \cdots a_k, \\ 0 & \text{otherwise.} \end{cases}$$

Thus, for $a = 11$, we have $C_a(x) = 1 + x$.

Theorem 4.5 *Let $f_a(n)$ be the number of binary strings of length n excluding the substring a of length k. Then the generating function $F_a(x) = \sum_{n \geq 0} f_a(n) x^n$ is given by*

$$F_a(x) = \frac{C_a(x)}{x^k + (1 - 2x)C_a(x)},$$

where $C_a(x)$ is the correlation polynomial of a.

Proof We define $g_a(n)$ to be the number of binary sequences of length n which commence with a but have no other occurrence of a as a consecutive subsequence, and $G_a(x) = \sum_{n \geq 0} g_a(n) x^n$ the generating function of this sequence of numbers.

Let b be a sequence counted by $f_a(n)$. Then for $x \in \{0, 1\}$, the sequence xb contains a at most once at the beginning. So

$$2 f_a(n) = f_a(n+1) + g_a(n+1).$$

Multiplying by x^n and summing over $n \geq 0$ gives

$$2 F_a(x) = x^{-1}(F_a(x) - 1 + G_a(x)). \tag{4.2}$$

Now let c be the concatenation ab. Then c starts with a, and may contain other occurrences of a, but only at positions overlapping the initial a, that is, where $a_{k-j+1} \cdots a_k b_1 \cdots b_{k-j} = a_1 \cdots a_k$. This can only occur when $c_a(k-j) = 1$, and the sequence $a_{k-j+1} \cdots a_k b$ then has length $n + j$ and has a unique occurrence of a at the beginning. So

$$f_a(n) = \sum g_a(n+j),$$

where the sum is over all j with $1 \leq j \leq k$ for which $c_a(k-j) = 1$. This can be rewritten

$$f_a(n) = \sum_{j=1}^{k} c_a(k-j) g_a(n+j),$$

or in terms of generating functions,

$$F_a(x) = x^{-k} C_a(x) G_a(x). \tag{4.3}$$

Combining equations (4.2) and (4.3) gives the result.

In the case where $a = 11$, we obtain

$$F_{11}(x) = \frac{1+x}{x^2 + (1-2x)(1+x)} = \frac{1+x}{1-x-x^2},$$

so that $f_{11}(n) = F_n + F_{n-1} = F_{n+1}$, as previously noted.

4.1.3   Horizontally convex polyominoes

I am grateful to Thomas Müller for suggesting this example.

A *polyomino* with n cells (sometimes called an n-omino) is a figure formed from n unit squares, which is connected in the sense that we can move from any square to any other by a path which crosses edges of the squares not at a corner. A polyomino is *horizontally convex* or HC if the cells in any row are all contiguous, with no gaps. So a 3×3 block of unit squares is HC, but if we left out the centre square it would not be.

We are going to count the HC polyominoes. The argument is quite long and the manipulations are not written out in detail.

Let a_n be the number of HC polyominoes made up of n cells, and let $f(x)$ be its generating power series $\sum a_n x^n$. We make the convention here that $a_0 = 0$, so that the sum starts at $n = 1$. The sequence begins

$$1, 2, 6, 19, 61, 196, 629, 2017, 6466, 20727, 66441, 212980, 682721, \ldots$$

In order to do the count, we introduce another parameter. Let $b_{m,n}$ be the number of HC polyominoes with n cells for which the bottom row consists of m cells. Thus we have

$$a_n = \sum_{m=1}^{n} b_{n,m},$$

with $b_{n,n} = 1$ and $b_{m,n} = 0$ if $m > n$. We let $g(x) = \sum \sum m b_{m,n}$. (Here and later, the double sum means over m and n, and the indices begin at 1.)

The crucial fact is that we can write down a recurrence relation for $b_{m,n}$. If we remove the bottom layer of a polyomino counted by $b_{m,n}$, with $m < n$, we obtain one counted by $b_{i,n-m}$, for some i. There are $m + i - 1$ ways of putting the bottom layer back, since the two layers must overlap. So

$$b_{m,n} = \sum_{i \geq 1} (m + i - 1) b_{i,n-m}$$

for $m < n$.

We take the two-variable generating function

$$F(x, y) = \sum \sum b_{m,n} x^n y^m.$$

We note that $f(x) = F(x, 1)$, while $g(x) = [(\partial/\partial y) F(x, y)]_{y=1}$.

We separate the expression for $F(x, y)$ into two parts: terms with $n = m$, and terms with $n \neq m$ (necessarily $n > m$). The first part is just

$$xy + x^2 y^2 + \cdots = \frac{xy}{1 - xy}.$$

The second part, which we shall call Σ, is

$$\sum_{m < n} b_{m,n} x^n y^m = \sum_{m < n} \sum_{i} (m + i - 1) b_{i,n-m}.$$

We write $\Sigma = \Sigma_1 + \Sigma_2 - \Sigma_3$, where each term is a sum over m, n, i with $m < n$, and the summands are $mb_{i,n-m}x^n y^m$, $ib_{i,n-m}x^n y^m$, and $b_{i,n-m}x^n y^m$ respectively. We have

$$
\begin{aligned}
\Sigma_1 &= \sum_{m\geq 1} m(xy)^m \sum_{n>m} x^{n-m} \sum_{i\geq 1} b_{i,n-m} \\
&= \sum_{m\geq 1} m(xy)^m \sum_{n>m} a_{n-m} x^{n-m} \\
&= f(x)\frac{xy}{(1-xy)^2},
\end{aligned}
$$

using the Binomial Theorem with exponent -2. Also

$$
\begin{aligned}
\Sigma_2 &= \sum_{m\geq 1} (xy)^m \sum_{n,i} ib_{i,n-m} x^{n-m} \\
&= g(x)\frac{xy}{1-xy},
\end{aligned}
$$

and

$$
\begin{aligned}
\Sigma_3 &= \sum_{m\geq 1} (xy)^m \sum_{n,i} b_{i,n-m} x^{n-m} \\
&= f(x)\frac{xy}{1-xy}.
\end{aligned}
$$

So we get

$$
F(x,y) = \frac{xy}{1-xy} + \left(\frac{xy}{1-xy}\right)^2 f(x) + \frac{xy}{1-xy} g(x).
$$

Putting $y = 1$, we obtain

$$
f(x) = \frac{x}{1-x} + \left(\frac{x}{1-x}\right)^2 f(x) + \frac{x}{1-x} g(x).
$$

Differentiating and putting $y = 1$, we obtain

$$
g(x) = \frac{x}{(1-x)^2} + \frac{2x^2}{(1-x)^3} f(x) + \frac{x}{(1-x)^2} g(x).
$$

We now have two equations for $f(x)$ and $g(x)$, which we can solve to obtain

$$
f(x) = \frac{x(1-x)^3}{1 - 5x + 7x^2 - 4x^3}.
$$

By reversing the proof of Theorem 4.2, we conclude that the number a_n of HC polyominoes with n cells satisfies the recurrence relation

$$
a_n = 5a_{n-1} - 7a_{n-2} + 4a_{n-3}
$$

for $n \geq 5$. (Remember from the proof that the recurrence relation holds for n at least one greater than the degree of the polynomial in the numerator.)

It is worth spending a little while trying to deduce this recurrence relation directly from the definition of a_n. You will probably not succeed; no such direct proof is known.

4.1.4 C-finite sequences

This section is based on a survey article by Doron Zeilberger [20]. C-finite sequences are those which satisfy recurrence relations with constant coefficients. They have a number of remarkable properties, which yield a method for showing that two sequences are equal, by a finite amount of computation. We illustrate with one of Zeilberger's examples.

Let $a = (a_n)$ be a sequence of rational numbers. We define the *shift* σa by moving each term one place left, deleting the first term. Then for a positive integer k, $\sigma^k a$ denotes the sequence obtained by shifting k times, with $\sigma^0 a = a$.

Proposition 4.6 *The following conditions on the sequence a are equivalent:*

(a) *a satisfies a recurrence relation with constant coefficients;*

(b) *the generating function $f(x)$ of a is a rational function (the quotient of two polynomials), where the denominator has non-zero constant term;*

(c) *there is a finite-dimensional subspace of the space of all rational sequences which contains a and all of its shifts.*

Proof We have seen the equivalence of (a) and (b) in Theorem 4.2.

If (a) holds, where the recurrence has $d + 1$ terms (that is, has the form

$$a_n = \sum_{i=1}^{d} \alpha_i a_{n-i}$$

for $n \geq d$), then $\sigma^d a = \sum_{i=1}^{d} \alpha_i \sigma^{d-i} a$, so that $\sigma^d a$ lies in the span of the sequences $a, \sigma a, \ldots, \sigma^{d-1} a$. By an easy induction, $\sigma^n a$ lies in this subspace for all $n \geq d$, and (c) holds.

Conversely, if (c) holds, then we may assume that the subspace is spanned by finitely many shifts of a. So $\sigma^d a$ is a linear combination of $a, \sigma a, \ldots, \sigma^{d-1} a$ for some d, and a satisfies a $(d + 1)$-term recurrence with constant coefficients.

We denote the subspace spanned by the shifts of a by $S(a)$.

A *C-finite sequence* is one which satisfies the three equivalent conditions of the proposition. Its *degree* can be defined as the smallest d such that the sequence satisfies a $(d + 1)$-term recurrence; the smallest degree of the denominator of a rational function representing the generating function; or the dimension of the vector space spanned by the shifts of a.

From this, we see immediately:

Proposition 4.7 *The sum of two C-finite sequences is C-finite; its degree is at most the sum of the degrees of the two sequences.*

Proof Clearly $a, b \in S(a) + S(b)$ (where the sum is taken in the space of all sequences); and so $a + b \in S(a) + S(b)$, a finite-dimensional vector space. Clearly all shifts of $a + b$ also lie in this space.

Note that the C-finite sequences form a vector space over $\mathbb{Q}$, since scalar multiplication clearly takes a C-finite sequence to one of the same degree.

Corollary 4.8 *Let a and b be two C-finite sequences with degrees d and e respectively, and suppose that the first $d + e$ terms of a and b agree. Then $a = b$.*

Proof $a - b$ is a C-finite sequence of degree at most $d + e$ whose first $d + e$ terms are zero; so $a - b$ is identically zero.

The power of this test is increased by the following result. The *pointwise product* $a \circ b$ of two sequences $a = (a_n)$ and $b = (b_n)$ is the sequence whose nth term is $a_n b_n$.

Proposition 4.9 (a) *Let a and b be two C-finite sequences with degrees d and e respectively. Then $a \circ b$ is a C-finite sequence with degree at most de.*

(b) *Let a be a C-finite sequence with degree d. Then $a \circ a$ is a C-finite sequence with degree at most $d(d + 1)/2$.*

(c) *More generally, if a is C-finite with degree d, then the sequence of k-th powers of elements of a is C-finite with degree at most $\dbinom{d + k - 1}{k}$.*

Proof (a) We consider the set of all infinite matrices (with rows and columns indexed by the natural numbers) such that each row satisfies the $(d + 1)$-term recurrence defining a and each column satisfies the $(e + 1)$-term recurrence defining b. It is easy to see that this set is a rational vector space. Moreover, any such matrix is completely determined by the de entries in the first e rows and d columns, so the vector space has dimension de.

A particular matrix lying in this space has entry $b_i a_j$ in row i and column j. All its simultaneous shifts (move everything one place up and to the left, deleting the first row and column) also belong to this space. Now consider the linear map which 'projects' the vector space onto the diagonal (that is, it maps any matrix to the sequence of entries on the diagonal). This map is linear, and so its dimension is at most de. Projecting the matrix with entries $b_i a_j$ and its shifts by the same horizontal and vertical amount onto the diagonal, we see that the product sequence and all its shifts are contained in a space of dimension at most de. The result follows.

(b) The proof is similar, except that we can use the space of symmetric matrices whose rows and columns satisfy the recurrence. To specify such a matrix, we only need to give the $d(d+1)/2$ entries on and above the diagonal in the top left $d \times d$ square.

(c) In the argument in part (b) above, instead of using a matrix, which is a '2-dimensional array', we use a 'k-dimensional array'. The space of completely symmetric k-dimensional arrays of side d has dimension equal to the number of selections of k things from $\{1, \ldots, d\}$ with repetition allowed and order unimportant.

Here is another closure property of C-finite sequences.

Proposition 4.10 *Let $a = (a_n)$ be a C-finite sequence of degree d. Then the sum sequence $s = (s_n)$, where $s_n = \sum_{i=0}^{n} a_i$, is C-finite of degree at most $d+1$.*

Proof Let $F(x)$ be the generating function of a, a rational function whose denominator has degree d (and non-zero constant term). The sum sequence s is the convolution of a with the all-1 sequence, whose generating function is $1/(1-x)$; so the generating function for s is $f(x)/(1-x)$, a rational function with denominator of degree at most $d+1$.

We can use our knowledge of C-finiteness to establish formulae for sequences with a finite amount of computation. Here is a simple example.

Example The sum of the first n Fibonacci numbers is $F_{n+2} - 1$.

To prove this, we observe that the Fibonacci sequence is C-finite of degree 2, so its sum sequence is C-finite of degree at most 3; also the right-hand side is C-finite of degree at most 3 (it is the sum of the Fibonacci sequence, of degree 2, and a constant sequence, of degree 1). So to prove the assertion, it suffices to verify that the first six values of the two sequences agree.

In this case it is not too hard to prove the result directly; the point is that, in the above argument, no thinking is required, the verification is purely mechanical.

Example: Domino tiling of a frame. Donald Knuth, in the *Mathematics Magazine*, posed the problem: For $n \geq 2$, consider the figure obtained from a square of side $n+2$ by removing a square of side $n-2$ with the same centre and sides parallel to those of the original square. In how many ways can this figure be tiled with 1×2 dominoes, which can be placed in either orientation?

Zeilberger, in the cited article, describes this example in detail. The answer is $4(2F_n^2 + (-1)^n)^2$, where F_n is the nth Fibonacci number. Several people gave an ingenious proof of this. However, once the formula has been guessed, and it has been checked that both sides of the equality are C-finite sequences (and bounds on their degree established), the result can be checked by mechanical computation (or, better, by computation involving some clever tricks to reduce the work required).

I won't give the proof in detail. The fact that the number of domino tilings is C-finite is obtained using the *transfer matrix method*. In essence, there are only a finite number of ways of covering the four 4×4 corner squares with dominoes, some of which may stick out. No matter what corner tilings are chosen, the number of ways of tiling the sides satisfy the same recurrence relation, with initial conditions depending on the corner tilings chosen. So the required number is a sum of fourth powers of sequence satisfying a 3-term recurrence (hence of degree 2), and so itself has degree 5, by Proposition 4.9(c).

The fact that the conjectured answer is C-finite follows from general results about C-finite sequences. The Fibonacci sequence has dimension 2, so its square has dimension 3; the sequence $(-1)^n$ has dimension 1, so the sum has dimension at most 4, and its square has dimension at most 10. So the result can be proved by checking 15 values.

However, with some ingenuity, we can do better. Write the conjectured answer as $16F_n^4 + 16(-1)^n F_n^2 + 4$. Now the fourth powers of Fibonacci numbers have dimension 5, and the recurrence relation can be computed explicitly; then it can be shown that the sequence $((-1)^n F_n^2)$ and the constant sequence both satisfy the same recurrence relation. So the dimension is 5, and only 10 terms need to be checked, a substantial saving.

4.1.5   Unbounded recurrences

The definition of a recurrence relation allows the value of the nth term u_n to depend on all the preceding values. We saw at the start of this chapter an example, the number u_n of compositions of the integer n. We showed that this satisfies the recurrence relation

$$u_n = 1 + u_1 + u_2 + \cdots + u_{n-1},$$

which we were able to reduce to the much simpler recurrence

$$u_1 = 1, \qquad u_n = 2u_{n-1} \text{ for } n > 1.$$

We will give here just one further example, a very important one. Recall from the last chapter that the generating function for the number $p(n)$ of partitions of the integer n is given by

$$\sum_{n \geq 0} p(n)x^n = \left(\prod_{k \geq 1} (1 - x^k) \right)^{-1}.$$

Thus, to get a recurrence relation for $p(n)$, we have to understand the coefficients of its inverse:

$$\sum_{n \geq 0} a(n)x^n = \prod_{k \geq 1} (1 - x^k).$$

Now a term on the right arises from each expression for n as a sum of distinct positive integers; its value is $(-1)^k$, where k is the number of terms in the sum. Thus, $a(n)$ is equal to the number of expressions for n as the sum of an even number of distinct parts, minus the number of expressions for n as the sum of an odd number of distinct parts.

This number is evaluated by *Euler's Pentagonal Numbers Theorem*:

Proposition 4.11

$$a(n) = \begin{cases} (-1)^k & \text{if } n = k(3k-1)/2 \text{ for some } k \in \mathbb{Z}, \\ 0 & \text{otherwise.} \end{cases}$$

We will prove this theorem in an appendix to this chapter.

Putting all this together, the recurrence relation for $p(n)$ is

$$\begin{aligned} p(n) &= \sum_{k \neq 0} (-1)^{k-1} p(n - k(3k-1)/2) \\ &= p(n-1) + p(n-2) - p(n-5) - p(n-7) + p(n-12) + \cdots \end{aligned}$$

where the summation is over all values of k for which $n - k(3k-1)/2$ is non-negative.

The number of terms in the recurrence grows with n, but only like $c\sqrt{n}$. So evaluating $p(n)$ for $n \leq N$ requires only $cn^{3/2}$ additions and subtractions, for some constant c.

4.2 Linear recurrence relations with polynomial coefficients

There is a similar, but less complete, theory of linear recurrence relations with polynomial coefficients. In this case, rather than a simple expression for the generating function as a rational function (a quotient of two polynomials), we obtain a linear differential equation. Sequences with this property are called *D-finite*.

We begin with a couple of examples, and then discuss a 'general' method (which, however, does not always work, unlike the case of constant coefficients).

4.2.1 Derangements

Let $d(n)$ be the number of derangements of $\{1, \ldots, n\}$ (permutations which have no fixed points). We obtain a recurrence relation as follows. Each derangement maps n to some i with $1 \leq i \leq n-1$, and by symmetry each i occurs equally often. So we need only count the derangements mapping n to $n-1$, and multiply by $n-1$.

We divide these derangements into two classes. The first type map $n-1$ back to n. Such a permutation must be a derangement of $\{1, \ldots, n-2\}$ composed with the transposition $(n-1, n)$; so there are $d(n-2)$ such. The second type map i to n for some $i \neq n-1$. Replacing the sequence $i \mapsto n \mapsto n-1$ by the sequence $i \mapsto n-1$, we obtain a derangement of $n-1$; every such derangement arises. So there are $d(n-1)$ derangements of this type.

Thus,

$$d(n) = (n-1)(d(n-1) + d(n-2)).$$

This is a linear recurrence relation, where the coefficients, rather than being constants, are polynomials in n.

There is a simpler recurrence satisfied by $d(n)$, which can be deduced from this one, namely

$$d(n) = nd(n-1) + (-1)^n.$$

To prove this by induction, suppose that it is true for $n-1$. Then $(n-1)d(n-2) = d(n-1) - (-1)^{n-1}$; so $d(n) = (n-1)d(n-1) + d(n-1) + (-1)^n$, and the inductive step is proved. (Starting the induction is an exercise.)

Now this is a special case of a general recursion which can be solved, namely

$$u_0 = c, \qquad u_n = p_n u_{n-1} + q_n \text{ for } n \geq 1.$$

We can include the initial condition in the recursion by setting $q_0 = c$ and adopting the convention that $u_{-1} = 0$.

If $q_n = 0$ for $n \geq 1$, then the solution is simply $u_n = P_n$ for all n, where

$$P_n = c \prod_{i=1}^{n} p_i.$$

So we compare u_n to p_n. Putting $y_n = u_n / P_n$, the recurrence becomes

$$y_0 = 1, \qquad y_n = y_{n-1} + \frac{q_n}{P_n} \text{ for } n \geq 1,$$

with solution

$$y_n = \sum_{i=0}^{n} \frac{q_i}{P_i}.$$

(Remember that $q_0 = P_0 = c$.) Finally,

$$u_n = P_n \sum_{i=0}^{n} \frac{q_i}{P_i}.$$

For derangements, we have $p_n = n$, $c = 1$ (so that $P_n = n!$), and $q_n = (-1)^n$. Thus

$$d(n) = n! \sum_{i=0}^{n} \frac{(-1)^i}{i!}.$$

It follows that $d(n)$ is the nearest integer to $n!/e$, since

$$n!/e - d(n) = n! \sum_{i \geq n+1} \frac{(-1)^i}{i!},$$

and the modulus of the alternating sum of decreasing terms on the right is smaller than that of the first term, which is $n!/(n+1)! = 1/(n+1)$.

The exponential generating function for the derangement numbers is

$$\sum_{n \geq 0} \frac{d(n)}{n!} x^n = \left(\sum_{i \geq 0} \frac{(-1)^i}{i!} x^i \right) \cdot \left(\sum_{j \geq 0} x^j \right) = \frac{e^{-x}}{1-x}.$$

4.2.2 Two generalisations

The count of derangements can be extended in various ways. I mention two here, without proof.

First extension: numbers of fixed points If a random permutation is chosen from the symmetric group of degree n, the probability that it is a derangement is very close to $1/e$. More generally, the number of fixed points of a random permutation is a random variable, taking non-negative integer values, and we can ask about its distribution. This turns out to be approximately a Poisson random variable with parameter 1 (and indeed it converges to this distribution as $n \to \infty$).

This means that, if k is small compared to n, the number of permutations with k fixed points is approximately $e^{-1}/k!$.

In Exercise 7.6 we will see how the generating function for these probabilities (for fixed n) can be computed.

More generally one could ask about the joint distribution of the numbers of cycles of lengths $1, 2, \ldots$. These are summarised in a multivariate polynomial which we will meet in Chapter 7. They are approximately independent Poisson random variables, where the variable for k-cycles has parameter $1/k$.

Second extension: derangements of k-sets Given n and k, how many permutations of $\{1, \ldots, n\}$ leave no k-element set fixed? This number tends to a limit as $n \to \infty$ for fixed k. The value of the limit is somewhat mysterious, but a fairly detailed account of its asymptotics (in terms of k) has been given by Eberhard, Ford and Green.

4.2.3 Involutions

Let s_n be the number of elements π of the symmetric group S_n which satisfy $\pi^2 = 1$ (this means that the cycles of π have lengths 1 or 2).

We have the recurrence relation

$$s_n = s_{n-1} + (n-1)s_{n-2}.$$

For count involutions $\pi \in S_n$ according to their effect on the last point n. If $n\pi = n$, then π is obtained from a unique involution in S_{n-1} by adjoining the fixed point n. If $n\pi = i \leq n-1$, then $i\pi = n$; and π is obtained from a unique involution in S_{n-2} by adjoining the 2-cycle (i,n). Since there are $n-1$ possible values of i, we have the result.

Using this recurrence, with initial values $s_0 = s_1 = 1$, it can be shown (see Exercise 4.12) that the exponential generating function of (s_n) is

$$\sum_{n \geq 0} \frac{s_n x^n}{n!} = \exp\left(x + \frac{x^2}{2}\right).$$

This is an example of an *exponential principle*, which we will discuss later. Basically, any involution can be uniquely decomposed into 'indecomposable' involutions, namely 1-cycles and 2-cycles. There is one of each type of indecomposable, so the e.g.f. for indecomposable involutions is simply $x + x^2/2$. Then the e.g.f. for arbitrary involutions is the exponential of this.

Another simple example of the exponential principle concerns arbitrary permutations. The 'indecomposable' permutations are cycles; there are $(n-1)!$ cycles of length n, and so the e.g.f. is

$$\sum_{n \geq 1} \frac{(n-1)! x^n}{n!} = -\log(1-x);$$

the exponential of this is

$$(1-x)^{-1} = \sum_{n \geq 0} \frac{n! x^n}{n!},$$

the e.g.f. for the sequence $(n!)$ counting all permutations.

4.2.4   Bell numbers

We already calculated the exponential generating function for the Bell numbers. Here is how to do it using the recurrence relation

$$B(n) = \sum_{k=1}^{n} \binom{n-1}{k-1} B(n-k).$$

Multiply by $x^n/n!$ and sum over n: the e.g.f $F(x)$ is given by

$$F(x) = \sum_{n \geq 0} \frac{x^n}{n!} \sum_{k=1}^{n} \binom{n-1}{k-1} B(n-k).$$

Differentiating with respect to x we obtain

$$\frac{d}{dx}F(x) = \sum_{n\geq 1}\frac{x^{n-1}}{(n-1)!}\sum_{k=1}^{n}\binom{n-1}{k-1}B(n-k)$$

$$= \sum_{l\geq 0}\frac{x^l}{l!}\sum_{m\geq 0}\frac{B(m)x^m}{m!}.$$

Here we use new variables $l = k - 1$ and $m = n - k$; the constraints of the original sum mean that l and m independently take all natural number values. Hence

$$\frac{d}{dx}F(x) = \exp(x)F(x).$$

This first-order differential equation can be solved in the usual way with the initial condition $F(0) = 1$ to give

$$F(x) = \exp(\exp(x) - 1),$$

in agreement with our earlier result.

4.2.5 A general method

There is a general approach to linear recurrences with polynomial factors, which we outline next. This approach leads to a differential equation satisfied by the generating function. Often this equation cannot be solved; even when it can, it usually does not lead directly to a formula for the terms in the sequence. Nevertheless, it can be useful in finding asymptotics.

The approach depends on the following facts. Let (a_n) be a sequence and $\sum_{n\geq 0}a_nx^n$ its generating function.

(a) $xf(x) = \sum_{n\geq 1}a_{n-1}x^n$ is the generating function of the shifted sequence.

(b) $x(d/dx)f(x) = \sum_{n\geq 0}na_nx^n$ (see Exercise 2.1). More generally, writing D_x for d/dx, for any polynomial p, the generating function for the sequence $p(n)a_n$ is $p(xD_x)f(x)$.

(We have to be careful since the operations of differentiating and multiplying by x do not commute. For example, $(xD_x)^2$ means: differentiate, multiply by x, differentiate, multiply by x. This is *not* the same as differentiating twice and multiplying by x^2.)

Consider a recurrence relation of the form

$$p(n)a_n + q(n)a_{n-1} + r(n)a_{n-2} = 0,$$

for example, where p, q, r are polynomials. If $f(x)$ is the generating function of the sequence (a_n), we see that the coefficient of x^n in

$$p(x\mathrm{D}_x)f(x) + q(x\mathrm{D}_x)(xf(x)) + r(x\mathrm{D}_x)(x^2 f(x))$$

is equal to zero for $n \geq 2$. Hence

$$p(x\mathrm{D}_x)f(x) + q(x\mathrm{D}_x)(xf(x)) + r(x\mathrm{D}_x)(x^2 f(x)) = Ax + B$$

for some constants A, B which can be deduced from the initial conditions. With luck, the solutions of this differential equation will lead to the generating function f for our problem. See Exercise 4.15 for an example of what can go wrong.

4.2.6 Frobenius numbers

We use the above method on the following problem: How many permutations $\pi \in S_n$ satisfy $\pi^m = 1$, for given n, m?

Let $s_n^{(m)}$ be this number. The required permutations have the property that every cycle has length dividing m. Suppose that the cycle containing the point n has length d. We can regard n as being the first point in the cycle, and can choose the remaining points in the cycle in $(n-1)(n-2)\cdots(n-d+1)$ ways. Then we have to choose a permutation of the remaining $n-d$ points, in $s_{n-d}^{(m)}$ ways. Hence we have

$$s_n^{(m)} = \sum_{d \mid m} (n-1)\cdots(n-d+1)s_{n-d}^{(m)}.$$

By convention, we take $s_0^{(m)} = 1$ and $s_n^{(m)} = 0$ for $n < 0$, so the recurrence relation holds for all $n > 0$.

Since we are counting permutations, we let $f(x)$ be the exponential generating function for the sequence; let $f(x) = \sum a_n x^n$, where $a_n = s_n^{(m)}/n!$. Dividing the above recurrence relation by $n!$, we find that a factor of $(n-1)!$ can be cancelled, giving

$$na_n = \sum_{d \mid m} a_d.$$

By the general method, we have

$$x\mathrm{D}_x f(x) = \left(\sum_{d \mid m} x^d\right) f(x),$$

so

$$\frac{\mathrm{d}}{\mathrm{d}x} f(x) = \left(\sum_{d \mid m} x^{d-1}\right) f(x),$$

which can be solved by standard methods to yield

$$f(x) = A \exp\left(\sum_{d\mid m} \frac{x^d}{d}\right),$$

and the initial condition gives $A = 1$.

This agrees with our earlier result for $m = 2$ (counting involutions in S_n).

4.3 Some non-linear recurrence relations

Our treatment here is even less systematic; no general methods exist for solving arbitrary recurrence relations. However, the first example, Catalan numbers, is of very great importance.

4.3.1 Catalan numbers

It is sometimes possible to use a recurrence relation to derive an algebraic or differential equation for a generating function for the sequence. If we are lucky, this equation can be solved, and the resulting function used to find the terms in the sequence.

The nth *Catalan number* C_n is the number of ways of bracketing a product of n terms, where we are not allowed to assume that the operation is associative or commutative. For example, for $n = 4$, there are five bracketings

$$(a\circ(b\circ(c\circ d))), (a\circ((b\circ c)\circ d)), ((a\circ b)\circ(c\circ d)), ((a\circ(b\circ c))\circ d), (((a\circ b)\circ c)\circ d),$$

so $C_4 = 5$.

Any bracketed product of n terms is of the form $(A \circ B)$, where A and B are bracketed products of i and $n - i$ terms respectively. So

$$C_n = \sum_{i=1}^{n-1} C_i C_{n-i} \text{ for } n \geq 2.$$

Putting $F(x) = \sum_{n\geq 1} C_n x^n$, the recurrence relation shows that F and F^2 agree in all coefficients except $n = 1$. Since $C_1 = 1$ we have $F = F^2 + x$, or $F^2 - F + x = 0$. Solving this equation gives

$$F(x) = \tfrac{1}{2}(1 \pm \sqrt{1 - 4x}).$$

Since $C_0 = 0$ by definition, we must take the negative sign here.

This expression gives us a rough estimate for C_n: the nearest singularity to the origin is a branchpoint at $1/4$, so C_n grows 'like' 4^n. However, we can get the solution explicitly.

From the Binomial Theorem, we have

$$F(x) = \tfrac{1}{2}\left(1 - \sum_{n\geq 0}\binom{1/2}{n}(-4)^n\right).$$

Hence

$$
\begin{aligned}
C_n &= -\frac{1}{2}\binom{1/2}{n}(-4)^n \\
&= \frac{1}{2}\cdot\frac{1}{2}\cdot\frac{1}{2}\cdot\frac{3}{2}\cdots\frac{2n-3}{2}\cdot\frac{2^{2n}}{n!} \\
&= \frac{1}{2^{n+1}}\cdot\frac{(2n-2)!}{2^{n-1}(n-1)!}\cdot\frac{2^{2n}}{n\cdot(n-1)!} \\
&= \frac{1}{n}\binom{2n-2}{n-1}.
\end{aligned}
$$

In an appendix to this chapter, we will have more to say about Catalan numbers.

Sometimes we cannot get an explicit solution, but can obtain some information about the growth rate of the sequence.

4.3.2 Wedderburn–Etherington numbers

Another interpretation of the Catalan number C_n, as we will see in the Appendix, is the number of rooted binary trees with n leaves, where 'left' and 'right' are distinguished.

If we do not distinguish left and right, we obtain the *Wedderburn–Etherington numbers* W_n.

Such a tree is determined by the choice of trees with i and $n-i$ leaves, but the order of the choice is unimportant. Thus by Theorem 3.6, if $i = n/2$, the number of trees is only $W_i(W_i+1)/2$, rather than W_i^2. For $i \neq n/2$, we simply halve the number. This gives the recurrence

$$
W_n = \begin{cases}
\dfrac{1}{2}\displaystyle\sum_{i=1}^{n-1} W_i W_{n-i} & \text{if } n \text{ is odd,} \\[4mm]
\dfrac{1}{2}\left(\displaystyle\sum_{i=1}^{n-1} W_i W_{n-i} + W_{n/2}\right) & \text{if } n \text{ is even.}
\end{cases}
$$

Thus, $F(x) = \sum W_n x^n$ satisfies

$$F(x) = x + \tfrac{1}{2}(F(x)^2 + F(x^2)).$$

This cannot be solved explicitly. We will obtain a rough estimate for the rate of growth. Later, we find more precise asymptotics.

We seek the nearest singularity to the origin. Since all coefficients are real and positive, this will be on the positive real axis. (If a power series with positive real coefficients converges at $z = r$, then it converges absolutely at any z with $|z| = r$.) Let s be the required point. Then $s < 1$, so $s^2 < s$; so $F(z^2)$ is analytic at $z = s$. Now write the equation as

$$F(z)^2 - 2F(z) + (F(z^2) + 2z) = 0,$$

with 'solution'

$$F(z) = 1 - \sqrt{1 - 2z - F(z^2)}$$

(taking the negative sign as before). Thus, s is the real positive solution of

$$F(s^2) = 1 - 2s.$$

Solving this equation numerically (using the fact that $F(s^2)$ is the sum of a convergent Taylor series and can be estimated from knowledge of a finite number of terms), we find that $s \approx 0.403\ldots$, so that W_n grows 'like' $(2.483\ldots)^n$.

We will find more precise asymptotics for W_n in the final chapter of the book.

4.4 Appendix: Euler's Pentagonal Numbers Theorem

A *pentagonal number* is a number of the form $k(3k-1)/2$ or $k(3k+1)/2$ for some non-negative number k. Alternatively, it is a number of the form $k(3k-1)/2$ for some (positive, negative, or zero) integer k. The second description is preferable, since it generates zero once only, whereas the first produces zero twice. The reason for the name is shown by the pictures of pentagonal numbers for small positive k.

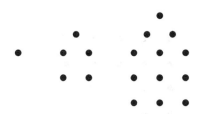

Figure 4.1: Small pentagonal numbers

The next theorem, due to Euler, is quite unexpected, as is its application: it will enable us to derive an efficient recurrence relation for the partition numbers.

Theorem 4.12 (Euler's Pentagonal Numbers Theorem)   *(a) If n is not a pentagonal number, then the numbers of partitions of n into an even and an odd number of distinct parts are equal.*

(b) *If* $n = k(3k-1)/2$ *for some* $k \in \mathbb{Z}$, *then the number of partitions of* n *into an even number of distinct parts exceeds the number of partitions into an odd number of distinct parts by one if* k *is even, and* vice versa *if* k *is odd.*

For example, if there are four partitions of $n = 6$ into distinct parts, viz. $6 = 5+1 = 4+2 = 3+2+1$, two of each parity; while if $n = 7$, there are five such partitions, viz. $7 = 6+1 = 5+2 = 4+3 = 4+2+1$, three with an even and two with an odd number of parts.

Proof   To demonstrate Euler's Theorem, we try to produce a bijection between partitions with an even and an odd number of distinct parts; we succeed unless n is a pentagonal number, in which case a unique partition is left out.

We represent a partition $\lambda : n = a_1 + a_2 + \cdots$, with $a_1 \geq a_2 \geq \cdots$, by a *diagram* $D(\lambda)$, whose ith row has i dots aligned on the left. See Figure 4.2 below for the diagram of the partition $17 = 6+5+4+2$. Such a diagram is essentially the same as a *Ferrers diagram* or *Young diagram*: we discuss these diagrams further in the next section. The only difference is that, instead of dots, these diagrams have boxes into which positive integers can be put.

Let λ be any partition of n into *distinct* parts. We define two subsets of the diagram $D(\lambda)$ as follows:

- The *base* is the bottom row of the diagram (the smallest part).

- The *slope* is the set of cells starting at the east end of the top row and proceeding in a south-westerly direction for as long as possible.

Note that any cell in the slope is the last in its row, since the row lengths are all distinct. See Figure 4.2.

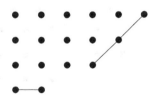

Figure 4.2: Base and slope

Now we divide the set of partitions of n with distinct parts into three classes, as follows:

- *Class 1* consists of the partitions for which *either* the base is longer than the slope and they don't intersect, *or* the base exceeds the slope by at least 2;

- *Class 2* consists of the partitions for which *either* the slope is at least as long as the base and they don't intersect, *or* the slope is strictly longer than the base;

- *Class 3* consists of all other partitions with distinct parts.

Given a partition λ in Class 1, we create a new partition λ' by removing the slope of λ' and installing it as a new base, to the south of the existing diagram. In other words, if the slope of λ contains k cells, we remove one from each of the largest k parts, and add a new (smallest) part of size k. This is a legal partition with all parts distinct. Moreover, the base of λ' is the slope of λ, while the slope of λ' is at least as large as the slope of λ, and strictly larger if it meets the base. So λ' is in Class 2.

In the other direction, let λ' be in Class 2. We define λ by removing the base of λ' and installing it as a new slope. Again, we have a partition with all parts distinct, and it lies in Class 1. (If the base and slope of λ meet, the base is one greater than the second-last row of λ', which is itself greater than the base of λ', which has become the slope of λ. If they don't meet, the argument is similar.)

The partition shown in Figure 4.2 is in Class 2; the corresponding Class 1 partition is shown in Figure 4.3.

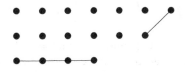

Figure 4.3: A Class 1 partition

These bijections are mutually inverse. Thus, the numbers of Class 1 and Class 2 partitions are equal. Moreover, these bijections change the number of parts by 1, and hence change its parity. So, in the union of Classes 1 and 2, the numbers of partitions with even and odd numbers of parts are equal.

Now we turn to Class 3. A partition in this class has the property that its base and slope intersect, and either their lengths are equal, or the base exceeds the slope by 1. So, if there are k parts, then $n = k^2 + k(k-1)/2 = k(3k-1)/2$ or $n = k(k+1) + k(k-1)/2 = k(3k+1)/2$. Figure 4.4 shows the two possibilities.

So, if n is not pentagonal, then Class 3 is empty; and, if $n = k(3k-1)/2$, for some $k \in \mathbb{Z}$, then it contains a single partition with $|k|$ parts. Euler's Theorem follows.

Corollary 4.13 $\displaystyle\prod_{n \geq 1}(1 - x^n) = \sum_{k=-\infty}^{\infty}(-1)^k x^{k(3k-1)/2}.$

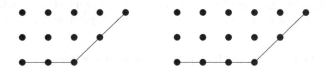

Figure 4.4: Two Class 3 partitions

Proof By Euler's Pentagonal Numbers Theorem, the right-hand side is the generating function for $\text{even}(n) - \text{odd}(n)$, where $\text{even}(n)$ and $\text{odd}(n)$ are the numbers of partitions having all parts distinct and having an even or odd number of parts respectively. We must show that the same is true for the left-hand side.

The coefficient of t^n is made up of contributions from factors of the form $(1 - x^{n_1}), \ldots, (1 - x^{n_k})$, where $n_1 + \ldots + n_k = n$ and $n_1, \ldots, n_k$ are distinct; the contribution from this choice of factors is $(-1)^k$. So each term counted by $\text{even}(n)$ contributes 1, and each term counted by $\text{odd}(n)$ contributes -1. So the theorem is proved.

The right-hand side can be written as

$$1 + \sum_{k>0} (-1)^k \left(x^{k(3k-1)/2} + x^{k(3k+1)/2} \right),$$

using the first 'definition' of the pentagonal numbers. From this, we deduce the promised recurrence for the partition numbers. This illustrates the general principle that finding a linear recurrence relation for a sequence is equivalent to finding the inverse of its generating function.

Corollary 4.14 *For $n > 0$,*

$$\begin{aligned}
p(n) &= \sum_{k>0} (-1)^{k-1} (p(n - \tfrac{1}{2}k(3k-1)) + p(n - \tfrac{1}{2}k(3k+1))) \\
&= p(n-1) + p(n-2) - p(n-5) - p(n-7) + p(n-12) + \ldots ;
\end{aligned}$$

with the convention that $p(n) = 0$ for $n < 0$.

Proof Since

$$\sum_{n\geq 0} p(n)x^n = \prod_{n>0}(1 - x^n)^{-1},$$

we have

$$\left(\sum_{n\geq 0} p(n)x^n \right) \cdot \left(1 + \sum_{k>0} (-1)^k (x^{k(3k-1)/2} + x^{k(3k+1)/2}) \right) = 1.$$

For $n > 0$, the coefficient of x^n in the product is zero. Thus,

$$0 = p(n) + \sum_{k>0} (-1)^k (p(n - \tfrac{1}{2}k(3k-1)) + p(n - \tfrac{1}{2}k(3k+1))),$$

from which the result follows.

This is a linear recurrence relation in which the number of terms grows with n, but relatively slowly: there are about $\sqrt{8n/3}$ pentagonal numbers below n. Thus, it permits efficient calculation: $p(n)$ can be evaluated with $cn^{3/2}$ additions or subtractions.

4.5 Appendix: Some Catalan objects

One of the remarkable features of the Catalan numbers is the very wide range of areas in which they arise. In this section we will look at a few of these. Not all details of proofs are given; filling these in is an exercise.

In particular, we will see several different classes of objects counted by Catalan numbers; there must therefore be bijections between these classes. In some cases, I will show, or outline, bijections; in other cases, it is an exercise (which can sometimes be solved by composing known bijections).

We begin by recalling that the Catalan number C_n is the number of ways of bracketing a non-associative product of n terms; it satisfies the recurrence

$$C_1 = 1, \quad C_n = \sum_{k=1}^{n-1} C_k C_{n-k} \text{ for } n \geq 2.$$

The ordinary generating fuction is

$$\sum_{n \geq 1} C_n x^n = 1 - \sqrt{1 - 4x},$$

and the formula for the Catalan numbers is

$$C_n = \frac{1}{n} \binom{2n-2}{n-1}.$$

4.5.1 Binary trees

Binary trees are always regarded as having a distinguished root; every vertex is either a leaf or has two descendants, which we distinguish as the left and right descendants. Thus, the following is a recursive specification:

- there is a binary tree consisting only of a root;
- any other binary tree T consists of an ordered pair (L, R) of binary trees (with their roots joined to the root of T).

Let T_n be the number of binary trees with n leaves. Then the first specification gives $T_1 = 1$, and the second gives

$$T_n = \sum_{k=1}^{n-1} T_k T_{n-k}$$

for $n \geq 2$. Clearly it follows that $T_n = C_n$ for all n.

In this case it is quite straightforward to match up the Catalan objects. Each bracketing of a product of n symbols has a 'parse tree'. We can regard this tree as being a machine for computing the product: at the leaves we input the elements to be multiplied; at each internal vertex we put a black box which performs the multiplication of two elements. Figure 4.5 shows the five binary trees with four leaves and the corresponding bracketed expressions.

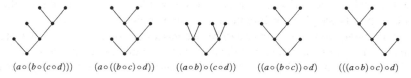

$(a \circ (b \circ (c \circ d)))$ $(a \circ ((b \circ c) \circ d))$ $((a \circ b) \circ (c \circ d))$ $((a \circ (b \circ c)) \circ d)$ $(((a \circ b) \circ c) \circ d)$

Figure 4.5: Binary trees and bracketed products

4.5.2   Rooted plane trees

Curiously, the Catalan numbers also count a much larger collection of trees! These are rooted plane trees which are not restricted to be binary. That is, a vertex of such a tree can have any number of descendants, these descendants being ordered. So the recursive specification is:

- there is a rooted plane tree consisting only of a root;

- any other rooted plane tree T consists of an ordered list of rooted plane trees $T_1, \ldots, T_k$ such that the roots of $T_1, \ldots, T_k$ are the descendants of the root of T.

Figure 4.6 gives the rooted plane trees with three edges.

Let R_n be the number of rooted plane trees with n edges. The first statement of the recursive specification shows that $R_0 = 1$. We divide trees T with at least one edge into two classes:

- those in which the root has only one descendant (clearly there are R_{n-1} of these, since each is just a rooted plane tree with $n - 1$ edges on top of a single-edge trunk);

Figure 4.6: Rooted plane trees

- those in which the root has more than one descendant. Suppose that the left-most descendant is a tree with k edges. There are R_k such trees. Each accounts for $k+1$ edges of the tree T (one of these is the edge through the root of T). Now deleting the left-most descendant gives an arbitrary rooted plane tree with $n - k - 1$ edges. Here k runs between 0 and $n - 1$.

So we have

$$R_n = \sum_{k=0}^{n-1} R_k R_{n-k-1}$$

for $n \geq 1$. Now it is easily proved by induction that $R_n = C_{n+1}$ for all n.

4.5.3 Dissections of polygons

A convex polygon with n sides can be dissected into triangles by drawing in $n - 2$ non-crossing diagonals. In how many ways can this be done? Figure 4.7 shows the five possible dissections of a pentagon.

Figure 4.7: Dissections of a polygon

Let D_n be the number of ways of dissecting an n-gon (with vertices numbered $0, 1, \ldots, n - 1$) into triangles.

Some of these dissections have the property that there is no diagonal containing the vertex 0. In this case, there must be a diagonal joining 1 to $n - 1$, and we have a dissection of the polygon with vertices $1, \ldots, n - 1$, with an extra triangle stuck on to the edge from 1 to $n - 1$. There are D_{n-1} ways to do this.

For any other dissection, suppose that k is minimal subject to the condition that there is a diagonal from 0 to k. This diagonal splits the given polygon into a $(k+1)$-gon with vertices $0, 1, \ldots, k$, and an $(n - k + 1)$-gon with vertices $0, k, k + 1, \ldots, n - 1$. The dissection of the first polygon has no diagonal containing 0, so there are D_k of these (using the preceding paragraph). The dissection of the second polygon is arbitrary, so there are D_{n-k+1} of these. Thus we have

$$D_n = D_{n-1} + \sum_{k=2}^{n-2} D_k D_{n-k+1},$$

where we take $D_2 = 1$.

From this it is an easy exercise to deduce that $D_n = C_{n-1}$ for $n \geq 2$.

4.5.4 Dyck paths

Catalan numbers also count certain types of paths joining lattice points in the plane. A *Dyck path* satisfies the following specifications:

- the path starts at the origin and ends at the point $(2n, 0)$;
- each step is either in the north-easterly direction (adding $(1, 1)$) or in the south-easterly direction (adding $(1, -1)$);
- the path never goes below the x-axis: that is, the y-coordinate is always non-negative.

Note that an even number of steps is required to return to the x-axis, since each step either increases or decreases the y-coordinate by 1. Figure 4.8 shows the paths for $n = 3$.

Figure 4.8: Dyck paths

Now we have the following result.

Proposition 4.15 *The number of Dyck paths from $(0,0)$ to $(2n,0)$ is the Catalan number C_{n+1}. Of these, the number which meet the x-axis only at their endpoints $(0,0)$ and $(2n,0)$ is C_n.*

Proof Let F_n be the total number of paths satisfying the specification, and G_n be the number which meet the axis only at the endpoints. By convention we take $F_0 = 1$. Figure 4.8 demonstrates that $F_3 = 5$ and $G_3 = 2$.

First we claim that $G_{n+1} = F_n$. For take any path satisfying the specification; translate it one step north-east, so that it goes from $(1, 1)$ to $(2n+1, 1)$; then extend it by a north-east step at the beginning and a south-east step at the end. The resulting path goes from $(0, 0)$ to $(2n+2, 0)$, and meets the axis only at its endpoints.

Conversely, given such a path, if we cut off its first and last steps and translate it by $(-1, -1)$, we obtain a path from $(0,0)$ to $(2n, 0)$ satisfying the original specification. So we have established a bijection between the sets counted by F_n and G_{n+1}.

Now take an arbitrary path counted by F_n, and suppose that its first return to the x-axis is at the point $(2k, 0)$. There are G_k ways to choose the part of the path before the point $(2k, 0)$, since it meets the axis only at its endpoints. Then there are F_{n-k} ways to choose the remaining part of the path, since it is just a path from $(0, 0)$ to $(2n - 2k, 0)$ satisfying the original specifications and moved east by $2k$. So

$$F_n = \sum_{k=1}^{n} G_k F_{n-k}$$

for $n \geq 1$.

From this it is easily deduced that $F_n = C_{n+1}$ and $G_n = C_n$.

4.5.5 A ballot problem

Here is another interpretation of Dyck paths.

Suppose that an election is held with two candidates, each of whom receives exactly n votes, so that the election is tied. In how many ways can the votes be counted so that candidate A is never behind? In how many ways is candidate A always strictly ahead except at the beginning and end of the count?

Given any way of counting the votes, we can represent it as a Dyck path in which a vote for A corresponds to a north-east step and a vote for B to a south-east step. The x-coordinate of the current point is the number of votes so far counted, and the y-coordinate is the number of votes by which A is ahead of B (so that negative y-coordinate corresponds to B being ahead in the count). The condition that the result is tied and A is never behind corresponds to the specification of Dyck paths in the last subsection, while counts where A is always ahead except at the beginning and end correspond to paths which meet the x-axis only at their ends. So the answers to the two questions above are the Catalan numbers C_{n+1} and C_n respectively.

For example, the counts corresponding to the five Dyck paths of Figure 4.8 are AAABBB, AABABB, AABBAB, ABAABB, and ABABAB respectively.

We will meet another class of Catalan objects in the last section of this chapter.

4.5.6 Tableaux

This analysis can be translated into a different language again, one of great importance in algebraic combinatorics.

Let us record the results of the count in the last subsection in a different way: take two rows corresponding to the two candidates, and as each vote is counted, enter its number in the row corresponding to the candidate for whom that vote is cast. The representations of the five counts above are shown in Figure 4.9.

AAABBB   AABABB   AABBAB   ABAABB   ABABAB

Figure 4.9: Tableaux

We see that Figure 4.9 shows all possible ways of putting the numbers $1, \ldots, 6$ into six boxes in a 2×3 array such that

- the numbers in each row are increasing (this corresponds to the fact that the votes are entered in order);
- the numbers in each column are increasing (this corresponds to the fact that B is never ahead of A in the count).

Such an arrangement is called a *tableau*. We can generalise by asking for arbitrary arrangements of boxes.

Let λ be a *partition* of the integer n: that is, λ is a sequence of positive integers, arranged in non-increasing order, with sum n. (The number of partitions of n is the *partition function* $p(n)$ we met earlier in this chapter.)

The *Young diagram*, or *Ferrers diagram*, associated with λ is an arrangement of boxes or cells in rows, the number of boxes in the ith row being the ith part of λ. Figure 4.10 shows the Young diagrams corresponding to the five partitions of 4.

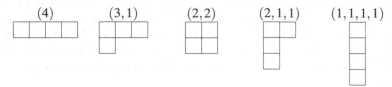

Figure 4.10: Young diagrams

In general, let λ be a partition of n, and $D(\lambda)$ the corresponding Young diagram. Because the parts are non-increasing, each row and column is a continuous run of boxes starting at the left or top edge. A *tableau*, or *Young tableau*, of *shape* λ, is an assignment of the numbers $1, \ldots, n$ to the cells of $D(\lambda)$ in such a way that the numbers in any row or column are increasing from the left or top edge of the diagram to the end of the row or column. We denote by f_λ the number of tableaux with shape λ.

Thus, the figures recording the results of a vote count of the type described in the preceding subsection, with two candidates obtaining m votes each, are precisely the tableaux of shape (m, m); so we have

$$f_{(m,m)} = C_{m+1} = \frac{1}{m+1}\binom{2m}{m}.$$

Arbitrary tableaux have also an interpretation in terms of ballots. Suppose that there are r candidates in an election with n voters, where the ith candidate receives m_i votes for $i = 1, \ldots, r$; we may arrange the candidates so that their totals are in non-increasing order (so that $\lambda = (m_1, \ldots, m_r)$ is a partition of n). Then f_λ is the number of ways of counting the votes if the candidates' totals at any stage are (non-strictly) in the order of the final result.

The numbers f_λ are of very great importance in algebra. Their most important interpretation is as follows. Let G be the symmetric group S_n of all permutations of $\{1, \ldots, n\}$. A *representation* of G is a homomorphism ρ from G to the group of non-singular linear transformations of a vector space V over the complex numbers. A representation is *irreducible* if the only subspaces of V mapped to themselves by all the matrices $\rho(g)$ are V itself and the zero space. Two representations are *equivalent* if they are related by a change of basis of V; that is, if $\rho_2(g) = T^{-1}\rho_1(g)T$ for some invertible transformation T.

It is known that any finite group has only finitely many irreducible representations up to equivalence; indeed, the number of irreducible representations is equal to the number of conjugacy classes of the group. Moreover, the degree of any such representation (the dimension of the vector space) divides the order of the group, and the sum of squares of the degrees of all irreducible representations is equal to the order of the group.

In the case where $G = S_n$, we have seen that two permutations are conjugate if and only if they have the same cycle structure. So the number of conjugacy classes, and hence the number of irreducible representations, is equal to $p(n)$. More is true in this case. The irreducible representations can be indexed by partitions of n; the degree of the representation indexed by λ is f_λ. Hence we conclude that

- f_λ divides $n!$;
- $\sum_\lambda f_\lambda^2 = n!$.

Both of these facts have a combinatorial interpretation. The first is the remarkable *hook length formula* for f_λ. For each cell in a Young diagram, the corresponding *hook* consists of this cell together with all cells to the right in the same row, or below in the same column. The *hook length* is equal to the number of cells in the hook.

Proposition 4.16 *Let λ be a partition of n. Then f_λ is equal to $n!$ divided by the product of the hook lengths of all the cells in $D(\lambda)$.*

For example, consider the partition (m,m) of $n = 2m$. The hook lengths of the cells in the first row of the diagram are $m+1, m, m-1, \ldots, 2$, while those for the cells in the bottom row are $m, m-1, \ldots, 1$. So the Proposition gives

$$f_{(m,m)} = \frac{(2m)!}{(m+1)!\, m!},$$

in agreement with the fact that $f_{(m,m)} = C_{m+1}$ and our earlier formula for Catalan numbers.

The second fact is explained by the *Robinson–Schensted–Knuth algorithm*:

Theorem 4.17 *There is a bijection between the set of permutations of $\{1, \ldots, n\}$, and the set of ordered pairs of tableaux of the same shape with n cells. Under this bijection, if π corresponds to the pair (S, T) of tableaux, then π^{-1} corresponds to (T, S). Furthermore, the number of cells in the first row of the tableaux corresponding to g is equal to the length of the longest increasing subsequence of g.*

The last part of the theorem has a 'dual' statement: The number of cells in the first column of the tableaux (that is, the number of rows) is equal to the length of the longest decreasing subsequence of g.

We will outline the proof of this theorem in the next section. The theorem immediately gives us the following counting results:

Corollary 4.18 (a) $\displaystyle\sum_\lambda f_\lambda^2 = n!$, *where the sum is over the set of all partitions of n.*

 (b) $\displaystyle\sum_\lambda f_\lambda = s_n$ *is equal to the number of permutations π of $\{1, \ldots, n\}$ for which π^2 is the identity, with the same range of summation as in (a).*

 (c) *The number of permutations of $\{1, \ldots, n\}$ with largest increasing subsequence of length m is equal to $\displaystyle\sum_\lambda f_\lambda$, where the sum is over all partitions of n with largest part m.*

Proof (a) The number of pairs of tableaux of the same shape is $\sum f_\lambda^2$.

 (b) A permutation π satisfies $\pi^2 = 1$ (that is, $\pi = \pi^{-1}$) if and only if it corresponds to a pair (S, S) of tableaux with its entries equal.

 (c) This is clear for the same reason as (a).

Note that we found a recurrence relation for the sequence (s_n) in Section 4.2.

For example, the numbers of tableaux of the three possible shapes with $n = 3$ are 1, 2 and 1 respectively. So S_3 contains $1^2 + 2^2 + 1^2 = 3!$ elements, of which $1 + 2 + 1 = 4$ satisfy $g^2 = 1$; and the numbers with longest increasing subsequence of length $1, 2, 3$ are $1, 4, 1$ respectively. All this is easily checked by listing the permutations.

4.5.7 Permutation patterns

For this application, we regard a permutation as simply an ordered list containing the numbers $1, 2, \ldots, n$ each once, rather than a bijective mapping on the set $\{1, 2, \ldots, n\}$.

For $m \leq n$, let σ and π be permutations of length m and n respectively. We say that π *contains the pattern* σ if there are m elements in π whose relative order is the same as that of the elements of σ. For example, the underlined elements show that $2\underline{1}6\underline{4}5\underline{3}$ contains the pattern 132: the first underlined element is the smallest of the three, and the second is the largest.

A lot of attention has been paid to the question: how many permutations of length n *avoid* (that is, do not contain) a fixed pattern? Two permutations are called *Wilf-equivalent* if the numbers of permutations of length n avoiding each of them are equal, for all n. Two transformations on permutations are easily seen to preserve Wilf equivalence: writing the permutation in reverse order (so that a_i is replaced by a_{n+1-i}, or reflecting the permutation in a horizontal axis (so that j is replaced by $n + 1 - j$). For example, these transformations replace 132 by 231 and 312 respectively; the second transformation applied to 231 gives 213. So these four permutations of $\{1, 2, 3\}$ are Wilf-equivalent. Similarly the other two permutations, 123 and 321, are Wilf-equivalent.

Early on, a surprising fact was discovered:

Proposition 4.19 *All six permutations of* $\{1, 2, 3\}$ *are Wilf-equivalent; the number of permutations of length n avoiding any one of them is the Catalan number C_n.*

Proving this, and finding bijections between the corresponding sets of permutations, is a challenging exercise.

However, the difficulty of the problem increases very rapidly. It is not known how many permutations of length n avoid 1324; even asymptotic estimates for this number are unknown.

4.6 Appendix: The RSK algorithm

In this section, we outline a constructive proof of Theorem 4.17. We build a pair (S, T) of tableaux from a permutation π, which we take in 'passive form' $(a_1, \ldots, a_n)$, where a_i is the image of i under π.

The construction proceeds step by step. Before the first step, S and T are empty. At the start of the ith step, S is a 'partial tableau' with i cells (this means that its entries are distinct but not necessarily the first i natural numbers, and the rows and columns are strictly increasing); and T is a tableau with the same shape as S.

In step i, we add a new cell to the shape, and add entries a_i to S and i to T, in a manner to be described. We define a recursive subroutine INSERT, which puts an integer a in the jth row of a partial tableau T.

If a is greater than the last element of the jth row, then simply append it to this row. (If the jth row is empty, put a in the first position.)

Otherwise, let x be the smallest element of the jth row for which $a \not> x$. (Work from right to left until an element smaller than a is first reached; then we have gone one step too far.) 'Bump' x out of the jth row, replacing it with a; then INSERT x into the $(j+1)$st row.

Now we can describe the RSK algorithm.

Start with S and T empty.
For $i = 1, \ldots, n$, do the following:

- INSERT a_i into the first row of S. This causes a cascade of 'bumps', ending with a new cell being created and a number (not less than a_i) written into it.
- Now create a new cell in the same position in T and write i into it.

To verify this algorithm, there are several things to check. We must show that the property that S is a partial tableau and T a tableau of the same shape holds after each step in the algorithm. The fact that rows and columns are increasing is, for S, a consequence of the way INSERT works; for T, it is because i is greater than any element previously in the tableau. The key point is that the newly created cell doesn't violate the condition that the row lengths are non-increasing; that is, there should be a cell immediately above it. This is because the element 'bumped' is smaller than the element to the right of the position it is 'bumped' out of, and so it comes to rest to the left of this position.

At the end of the algorithm, we have two tableaux of the same shape.

We illustrate the algorithm with the permutation $(2, 3, 1)$.

- In the first stage, put 2 into the first row of S, and 1 into the first row of T:

$$S = 2, \qquad T = 1.$$

- In the second stage, put 3 at the end of the first row of s, and 2 in the same position in T:

$$S = 2 \quad 3, \qquad T = 1 \quad 2.$$

- At stage 3, 2 is 'bumped' by 1 into the second row, giving S and T as shown:

$$S = \begin{matrix} 1 & 3 \\ 2 & \end{matrix}, \qquad T = \begin{matrix} 1 & 2 \\ 3 & \end{matrix}.$$

The procedure can be reversed, to construct a permutation from a pair of standard tableaux of the same shape. To see this, note that we can locate n in the tableau T, and then reconstruct the cascade of 'bumps' required to move the corresponding element of S to that position; the insertion triggering this cascade is a_n. Working back in the same way, we recover the entire permutation. (A few worked examples make this clearer than a written explanation.)

The proofs of the statements about inverse permutations and longest increasing subsequences require careful analysis of the algorithm, and will not be given here.

There is a more general form of the RSK algorithm, which gives a bijection between matrices of non-negative integers with elements summing to n and pairs of 'semi-standard' Young tableaux of the same shape. (A tableau is *semi-standard* if its rows are non-decreasing – repetitions are allowed – and its columns are strictly increasing. The tableaux we defined earlier are called *standard*.)

Further information can be found in the books by Macdonald or Stanley in the bibliography.

4.7 Appendix: Some inverse semigroups

The material in this section was communicated to me by Abdullahi Umar.

A *partial permutation* on a set X is a bijection between subsets of X. We allow the subsets to be empty or to be the whole of X. We denote the domain and range of the partial permutation f by $\mathrm{Dom}(f)$ and $\mathrm{Ran}(f)$ respectively.

Partial permutations can be composed by defining the product wherever possible; that is,

$$(f \circ g)(x) = \begin{cases} f(g(x)) & \text{if } x \in \mathrm{Dom}(g) \text{ and } g(x) \in \mathrm{Dom}(f), \\ \text{undefined} & \text{otherwise.} \end{cases}$$

A set of partial permutations which is closed under composition and contains the identity permutation on X is an *inverse semigroup*.

Consider the following three conditions on a partial permutation f of $\{1, \ldots, n\}$:

Monotonic: if $x, y \in \mathrm{Dom}(f)$ and $x < y$, then $f(x) < f(y)$;

Decreasing: if $x \in \mathrm{Dom}(f)$ then $f(x) \leq x$;

Strictly decreasing: if $x \in \mathrm{Dom}(f)$ then $f(x) < x$.

The set of partial permutations satisfying any collection of these conditions is an inverse semigroup. We proceed to find the number of elements in each of these inverse semigroups.

Let $P(n)$ be the number of partial permutations on $\{1, \ldots, n\}$. Denote the numbers of permutations which are respectively monotonic, decreasing, or strictly decreasing by a subscript m, d or s; we also allow combinations of subscripts. There is no known closed formula for $P(n)$; but all the other numbers are familiar combinatorial coefficients:

Theorem 4.20   (a) $P(n) = \sum_{k=0}^{n} \binom{n}{k}^2 \cdot k!$.

(b) $P_m(n) = \binom{2n}{n}$.

(c) $P_d(n) = B_{n+1}$ and $P_s(n) = B_n$, where B_n is the nth Bell number.

(d) $P_{md}(n) = C_{n+1}$ and $P_{ms}(n) = C_n$, where C_n is the nth Catalan number.

Proof (a) There are $\binom{n}{k}$ choices for the domain and $\binom{n}{k}$ choices for the range of a partial permutation of cardinality k. Once the domain and range are chosen, there are $k!$ bijections between them. The formula follows.

(b) Argue as above. Once the domain and range are chosen, there is a unique monotonic bijection between them. So

$$P_m(n) = \sum_{k=0}^{n} \binom{n}{k}^2 = \binom{2n}{n},$$

by Proposition 3.5 (putting $m = k = n$).

(c) We show first that $P_d(n) = P_s(n+1)$. If f is a decreasing partial permutation on $\{1, \ldots, n\}$, then the map g given by $g(x+1) = f(x)$ whenever this is defined is a strictly decreasing partial permutation on $\{1, \ldots, n+1\}$. The argument reverses. This correspondence preserves the property of being monotonic, so also $P_{md}(n) = P_{ms}(n+1)$.

Now we select a decreasing bijection by first choosing its fixed points, and then choosing a strictly decreasing bijection on the remaining points. If there are k fixed points, then there are $P_s(n-k)$ ways to choose the strictly decreasing bijection. So we have

$$P_s(n+1) = P_d(n) = \sum_{k=0}^{n} \binom{n}{k} P_s(n-k).$$

Thus, $P_s(n)$ satisfies the same recurrence as the Bell number B_n (see Section 4.2.4), and we have

$$P_s(n) = B_n, \qquad P_d(n) = B_{n+1}.$$

(d) The preceding proof fails for monotonic decreasing maps, since such a map cannot jump over a fixed point. Instead, we encode a strictly decreasing map by a Catalan object.

Let f be monotonic and strictly decreasing on $\{1, \ldots, n\}$. We encode f by a sequence of length $2n$ in the alphabet consisting of two symbols A and B as follows. In positions $2i - 1$ and $2i$, we put

AB, if $i \notin \mathrm{Dom}(f)$ and $i \notin \mathrm{Ran}(f)$,

AA, if $i \notin \mathrm{Dom}(f)$ and $i \in \mathrm{Ran}(f)$,

BB, if $i \in \mathrm{Dom}(f)$ and $i \notin \mathrm{Ran}(f)$,

BA, if $i \in \mathrm{Dom}(f)$ and $i \in \mathrm{Ran}(f)$.

It can be shown that this gives a bijective correspondence between the set of such functions and the set of solutions to the ballot problem in Section 4.5.5 above. The proof is an exercise. (It is necessary to show that the resulting string has equally many As and Bs, but each initial substring has at least as many As as Bs; and that every string with these properties can be decoded to give a strictly decreasing monotone function. The proof that the correspondence is bijective is then straightforward.)

It follows that $P_{ms}(n) = C_n$ (the nth Catalan number), and from the remark in part (c), also $P_{md}(n) = C_{n+1}$.

Remark Abdullahi Umar has found that many other interesting counting sequences arise in calculating the orders of various inverse semigroups of partial permutations. Among these are Fibonacci, Stirling, Schröder, Euler, Lah and Narayana numbers. Details are given in a sequence of papers by A. Laradji and A. Umar.

4.8 Exercises

4.1 Show directly that, with the hypotheses of Theorem 4.4, the sequence $u_n = n^j \alpha_i^n$ satisfies the recurrence (4.1) for $0 \le j \le m_i - 1$. Deduce the statement of the Theorem.

4.2 Solve the recurrence relation $u_n = u_{n-2}$

(a) directly;

(b) using the method of Theorem 4.4.

4.3 Let A be a finite set of positive integers. Suppose that the currency of a certain country has A as the set of denominations. Prove that the number $f(n)$ of ways of paying a bill of n units, where coins are paid in order (as in putting them into a vending machine), has generating function $1/(1 - \sum_{a \in A} x^a)$.

Suppose that $A = \{1, 2, 5, 10\}$. Prove that $f(n) \sim c\, \alpha^n$ for some constants c and α, and estimate α.

What is the generating function for the number in the case when the order of the coins is not significant (as in tendering a handful of coins in a shop)?

4.4 Let a be a binary string of length k with correlation polynomial $C_a(x)$. A random binary sequence is obtained by tossing a fair coin, recording 1 for heads and 0 for tails. Let E_a be the expected number of coin tosses until the first occurrence of a as a consecutive substring. Prove that E_a is the sum, over n, of the probability that a doesn't occur in the first n terms of the sequence. Deduce that $E_a = 2^k C_a(1/2)$.

4.5 This exercise is due to Wilf, and illustrates his 'snake oil' method.

(a) Prove that

$$\sum_{n\geq 0} \binom{n+k}{2k} x^{n+k} = \frac{x^{2k}}{(1-x)^{2k+1}}.$$

(b) Let

$$a_n = \sum_{k=0}^{n} \binom{n+k}{2k} 2^{n-k}$$

for $n \geq 0$. Prove that the ordinary generating function for (a_n) is

$$\sum_{n\geq 0} a_n x^n = \frac{1-2x}{(1-x)(1-4x)},$$

and deduce that $a_n = (2^{2n+1}+1)/3$ for $n \geq 0$.

(c) Write down a linear recurrence relation with constant coefficients satisfied by the numbers a_n.

4.6 Find an asymptotic formula for the number of horizontally convex polyominoes with n cells.

4.7 (a) Let $f(n)$ be the number of expressions for the natural number n as an *ordered* sum of 2's and 3's. Prove that

$$f(0) = 1, \quad f(1) = 0, \quad f(2) = 1, \quad \text{and}$$
$$f(n) = f(n-2) + f(n-3) \text{ for } n \geq 3.$$

Hence or otherwise show that

$$\sum_{n\geq 0} f(n)x^n = \frac{1}{1-x^2-x^3}.$$

Prove that the polynomial equation $1 - x^2 - x^3 = 0$ has no repeated roots, and deduce that $f(n) \sim c\alpha^n$ for some $c, \alpha > 0$. Calculate these constants to four decimal places.

(b) Let $g(n)$ be the number of expressions for the natural number n as an *unordered* sum of 2's and 3's. Prove that

$$\sum_{n \geq 0} g(n)x^n = \frac{1}{(1-x^2)(1-x^3)},$$

and deduce that $g(n) = g(n-2) + g(n-3) - g(n-5)$ for $n \geq 5$. Hence or otherwise prove that

$$g(n) \sim n/6.$$

4.8 According to Proposition 4.9(b), the sequence whose nth term is F_n^2 (the square of the nth Fibonacci number) satisfies a four-term recurrence relation. Find such a relation.

4.9 Prove Proposition 4.9(c).

4.10 Prove that the number of permutations of $\{1, \ldots, n\}$ which have exactly k fixed points is equal to $(n!/k!)d(n-k)$, where d denotes the derangement number.

4.11 Let a_n be the number of strings that can be formed from n distinct letters (using each letter at most once, and including the empty string). Prove that

$$a_0 = 1, \qquad a_n = na_{n-1} + 1 \text{ for } n \geq 1,$$

and deduce that $a_n = \lfloor e\, n! \rfloor$. What is the exponential generating function for this sequence?

4.12 Prove from the recurrence relation for the number s_n of involutions in S_n that its exponential generating function is $\exp(x + x^2/2)$.

Hint: Let $S(x) = \sum s_n x^n / n!$. Show that

$$\frac{\mathrm{d}}{\mathrm{d}x} S(x) = (1+x)S(x)$$

and deduce that

$$S(x) = A\exp(x + x^2/2);$$

use the initial value $s_0 = 1$ to show that $A = 1$.

4.13 Let $t_n = s_n^2$, where s_n is as in the preceding question. Prove that

$$t_n = nt_{n-1} + n(n-1)t_{n-2} + (n-1)(n-2)^2 t_{n-3}.$$

Remark s_n is the number of homomorphisms from the cyclic group C_2 to S_n. Hence t_n is the number of homomorphisms from the free product $C_2 * C_2$ to S_n, since such a homomorphism may be defined independently on the two generators.

4.14 Recall that the *parity* of a permutation π of $\{1,\ldots,n\}$ is the parity of $n - c(\pi)$, where $c(\pi)$ is the number of cycles of π. Let u_n be the number of even permutations π of $\{1,\ldots,n\}$ which satisfy $\pi^2 = 1$. Prove that

$$u_n = u_{n-1} + (n-1)(s_{n-2} - u_{n-2})$$

for $n \geq 2$.

Let $U(x)$ be the exponential generating function of (u_n). Prove that

$$\frac{\mathrm{d}}{\mathrm{d}x} U(x) = (1-x)U(x) + xS(x),$$

where $S(x)$ is as in Exercise 4.12, and deduce that

$$U(x) = \tfrac{1}{2}\left(\exp(x + x^2/2) + \exp(x - x^2/2)\right).$$

4.15 Consider the recurrence relation

$$a_n = na_{n-1}, \qquad a_0 = 1.$$

Clearly the unique solution is $a_n = n!$. What goes wrong if we try to apply the method of Section 4.2.5 to this example?

4.16 Show that the number of tableaux of shape $(n-k,k)$ for $1 \leq k \leq n/2$ is

$$\binom{n}{k} - \binom{n}{k-1}.$$

4.17   (a) Prove that the inverse of the RSK algorithm produces a permutation π from each pair (S,T) of tableaux of the same shape.

(b) Prove that, if π is produced by (S,T), then π^{-1} is produced by (T,S).

(c) Prove that, if the RSK algorithm adds a new element a in position i at the end of the first row of S, then a is the smallest element which lies at the end of an increasing subsequence of length i of the permutation π, and conversely. Deduce that, on conclusion of the algorithm, the length of the first row of S (and T) is equal to the length of the longest increasing subsequence of π.

4.18   (a) Let π be a permutation of $\{1,\ldots,n\}$. Suppose that the length of the longest increasing subsequence of π is m. For $i = 1,\ldots,m$, let A_i be the subset of $\{1,\ldots,n\}$ consisting of positions for which the longest increasing subsequence of π ending in that position is i. Show that each A_i carries a decreasing subsequence of $\{1,\ldots,n\}$. Deduce that the longest decreasing subsequence of π has size at least $\lceil n/m \rceil$.

(b) Hence show that π has either an increasing or a decreasing subsequence of length at least $\lceil \sqrt{n} \rceil$.

(c) Using the fact that the lengths of the longest increasing and decreasing subsequences of π are respectively the number of columns and number of rows of the Young tableaux corresponding to π under the RSK algorithm, give an alternative proof of the above facts.

4.19 Let k and l be positive integers, and $n = kl$. How many permutations of $\{1,\ldots,n\}$ have the property that the longest increasing subsequence has length k and the longest decreasing subsequence has length l?

Find all such permutations in the case $k = l = 2$.

4.20 Let $P(n)$ be the number of partial permutations of the set $\{1,\ldots,n\}$. Prove that $P(n)/n!$ tends to infinity faster than any polynomial in n.

CHAPTER 5

The permanent

The permanent of a square matrix is defined by the same sum of products as the determinant, but omitting the signs. Much of enumerative combinatorics involves calculating or estimating the permanent. However, there are no simple techniques for this. In this chapter we define the permanent and prove some of the main results about it (except for the proof of van der Waerden's conjecture), and apply them to the enumeration of Latin squares.

5.1 Permanents and SDRs

There is a well-known formula for the determinant of an $n \times n$ matrix $A = (a_{ij})$:

$$\det(A) = \sum_{\pi \in S_n} \text{sgn}(\pi) a_{1\,1\pi} a_{2\,2\pi} \cdots a_{n\,n\pi},$$

where S_n is the symmetric group of degree n, and $\text{sgn}(\pi)$ is the sign of the permutation π. This is a beautiful theoretical formula, but very inefficient for calculating determinants in practice, since it is a sum of $n!$ terms; a matrix can be reduced to upper triangular form (and its determinant calculated) with at most n^3 arithmetic operations by *Gaussian elimination*.

A very similar formula defines the *permanent* of A:

$$\text{per}(A) = \sum_{\pi \in S_n} a_{1\,1\pi} a_{2\,2\pi} \cdots a_{n\,n\pi}.$$

This looks simpler since the signs are not present. Unfortunately, there are no short-cuts for calculating the permanent. No method substantially faster than evaluating the formula is known.

106

However, a great deal of enumerative combinatorics is concerned with evaluating, or estimating, permanents of matrices with entries zero and one. We now explain the connection.

Let $(A_1, \ldots, A_n)$ be an n-tuple of subsets of a set X. A *system of distinct representatives*, or *SDR*, is an n-tuple $(a_1, \ldots, a_n)$ of elements of X such that

(a) $a_i \neq a_j$ for $i \neq j$;

(b) $a_i \in A_i$ for all i.

(The first condition says that the elements are distinct; the second, that they are representatives of the corresponding sets.)

Now suppose that $|X| = n$, say $X = \{1, 2, \ldots, n\}$. The sets $A_1, \ldots, A_n$ can be represented by an $n \times n$ *incidence matrix* $P = (p_{ij})$, where

$$p_{ij} = \begin{cases} 1 & \text{if } j \in A_i, \\ 0 & \text{otherwise.} \end{cases}$$

Proposition 5.1 *Let P be the incidence matrix of $(A_1, \ldots, A_n)$. Then $\mathrm{per}(P)$ is equal to the number of SDRs of $(A_1, \ldots, A_n)$.*

Proof Each non-zero term in the permanent of P has the form $p_{1\,1\pi} \cdots p_{n\,n\pi}$, where $i\pi \in A_i$ for $i = 1, \ldots, n$; thus $(1\pi, \ldots, n\pi)$ is an SDR for the family. Conversely, each SDR gives a non-zero term (with value 1) in the sum.

5.2 Hall's Theorem

In this section, we suppose that $(A_1, \ldots, A_n)$ is a family of subsets of a set X. Suppose that the family has an SDR $(a_1, \ldots, a_n)$. For any subset I of $\{1, \ldots, n\}$, the $|I|$ elements a_i (for $i \in I$) are all distinct and belong to the set

$$A(I) = \bigcup_{i \in I} A_i.$$

So we have

$$|A(I)| \geq |I| \text{ for all } i \subseteq \{1, \ldots, n\}.$$

This is known as *Hall's condition*, after Philip Hall, who proved that it is also sufficient:

Theorem 5.2 (Hall's Theorem) *The family $(A_1, \ldots, A_n)$ of subsets of X has a SDR if and only if $|A(I)| \geq |I|$ for all $I \subseteq \{1, \ldots, n\}$.*

Proof We have seen the necessity of the condition; now we prove its sufficiency. There are several proofs, of which the one given here is one of the simplest.

The proof is by induction on n. For $n = 1$, Hall's condition says $|A_1| \geq 1$, so certainly an SDR exists. So suppose, inductively, that any family with fewer than n sets which satisfies Hall's condition has an SDR.

We say that a subset of $\{1, \ldots, n\}$ is *critical* if $|A(I)| = |I|$. We divide the proof into two cases.

Case 1 No subset of $\{1, \ldots, n\}$ except $\emptyset$ and possibly $\{1, \ldots, n\}$ is critical. This means that $|A(I)| > |I|$ for any such set I. Choose any element $a_n \in A_n$, and let $A'_i = A_i \setminus \{a_n\}$ for $i = 1, \ldots, n-1$. For any non-empty $I \subseteq \{1, \ldots, n-1\}$, we have

$$|A'(I)| \geq |A(I)| - 1 \geq |I|,$$

since I is not critical. By induction, the family $(A'_1, \ldots, A'_{n-1})$ has an SDR, say $(a_1, \ldots, a_{n-1})$. Clearly $a_i \neq a_n$ for $i < n$, so $(a_1, \ldots, a_n)$ is an SDR for the original family.

Case 2 There is a critical set $J \neq \emptyset, \{1, \ldots, n\}$. By induction, there is an SDR $(a_j : j \in J)$ for the family $(A_j : j \in J)$; it uses all the $|J|$ elements of $A(J)$. For $k \notin J$, set $A^*_k = A_k \setminus A(J)$. Then for $K \subseteq \{1, \ldots, n\} \setminus J$, we have

$$
\begin{aligned}
|A^*(K)| &= |A(K \cup J)| - |A(J)| \\
&\geq |K \cup J| - |J| \\
&= |J|,
\end{aligned}
$$

where the second inequality holds because $|A(J)| = |J|$. Combining the two SDRs gives an SDR for the whole family.

5.3   The van der Waerden conjecture

A square matrix P is said to be *doubly stochastic* if its entries are non-negative real numbers and all its row and column sums are 1.

If a matrix of real numbers has all entries non-negative and all row and column sums equal to $k > 0$, then it can be converted to a doubly stochastic matrix by dividing it by k. We will exploit this in the next section.

Why are such matrices called 'doubly stochastic'? Suppose that the entries p_{ij} of an $n \times n$ matrix are non-negative and all the row sums are equal to 1. Then we can regard P as describing a stochastic system which can be in any one of n possible states; at a given instant, it makes a transition between states, and its probability of jumping from state i to state j is P_{ij}. (The row sum condition just says that the probabilities of all outcomes sum to 1.) Such a matrix is called *stochastic*.

The condition of being doubly stochastic requires the same condition for column sums, though its probabilistic interpretation is less clear.

Proposition 5.3 *The permanent of a doubly stochastic matrix is non-zero.*

Proof Let $X = \{1, \ldots, n\}$, and for $i = 1, \ldots, n$, let

$$A_i = \{j \in X : p_{ij} > 0\}.$$

We claim that $(A_1, \ldots, A_n)$ satisfies Hall's condition.

Take $I \subseteq \{1, \ldots, n\}$ and let $|I| = i$. Then the sum of all entries in the rows with index in I is precisely i. But, for every $j \in A(I)$, the sum of entries with row index in I and column index j is at most 1, since the sum of all entries in column j is 1. So we must have $|A(I)| \geq i = |I|$.

By Hall's Theorem, there is an SDR, say $(a_1, \ldots, a_n)$, for $(A_1, \ldots, A_n)$. Now $p_{1a_1} p_{2a_2} \cdots p_{na_n}$ is a non-zero term in the permanent of P; since all terms are non-negative, we have $\mathrm{per}(P) > 0$.

The argument can be used to show a little more, a theorem of Birkhoff. A *permutation matrix* is a matrix of zeros and ones with exactly one 1 in each row or column. A permutation π is represented by a permutation matrix $P(\pi)$ with (i, j) entry

$$p_{ij} = \begin{cases} 1 & \text{if } i\pi = j, \\ 0 & \text{otherwise;} \end{cases}$$

and every permutation matrix represents a unique permutation.

Proposition 5.4 *A doubly stochastic matrix is a linear combination of permutation matrices with positive coefficients summing to 1.*

Proof The proof is by induction on the number of non-zero entries in P. The smallest possible number of entries is n, with equality if and only if P is a permutation matrix.

So suppose that the proposition holds for doubly stochastic matrices with fewer non-zero elements than P. Choose an SDR as in the preceding proof, let us say $(a_1, \ldots, a_n)$, and let

$$q = \min\{P_{1a_1}, \ldots, P_{na_n}\}.$$

Then $0 < q \leq 1$; and we may assume that $0 < q < 1$, since $q = 1$ only if P is a permutation matrix.

Let B be the permutation matrix corresponding to the permutation π with $i\pi = a_i$ for $i = 1, \ldots, n$. Then

$$P' = (P - qB)/(1 - q)$$

is a doubly stochastic matrix with fewer non-zero entries than P; so by induction, we have $P' = \sum x_i B_i$, where $x_i > 0$ and $\sum x_i = 1$. But then

$$P = qB + (1-q)\sum x_i B_i,$$

and $(1-q) + q\sum x_i = 1$, as required.

Can we improve the lower bound of zero for the permanent?

The matrix $(1/n)J$, where J has every entry 1, is doubly stochastic, and has permanent $n!/n^n$. It was conjectured by van der Waerden that this matrix is extremal; that is, any doubly stochastic matrix P of order n satisfies $\operatorname{per}(P) \geq n!/n^n$, with equality if and only if $P = (1/n)J$.

The conjecture was proved independently by Egorychev and Falikman in 1981. The proof is just a little too long to be given here. You will find an accessible account of it in the book by van Lint and Wilson (see the bibliography).

Theorem 5.5 (Egorychev–Falikman Theorem) *If P is a doubly stochastic matrix of order n, then $\operatorname{per}(P) \geq n!/n^n$, with equality if and only if $P = (1/n)J$.*

5.4   Latin squares

A *Latin square* of order n is an $n \times n$ array with entries taken from the set $\{1, \ldots, n\}$ of symbols, such that each symbol occurs once in each row and once in each column.

For example, the following are the two Latin squares of order 3 with first row $(1, 2, 3)$. Since the first row is arbitrary, there are $2 \times 6 = 12$ Latin squares of order 3 altogether.

1	2	3
2	3	1
3	1	2

1	2	3
3	1	2
2	3	1

Let $L(n)$ be the number of Latin squares of order n. As hinted in the Preface, the problem of calculating $L(n)$ exactly seems to be intractible. Values are known for $n \leq 11$, as a result of extensive computation. The first ten are given in the following table. (The last entry was found by Brendan McKay and Ian Wanless in 2005.)

n	$L(n)$
1	1
2	2
3	12
4	576
5	161280
6	812851200
7	61479419904000
8	108776032459082956800
9	5524751496156892842531225600
10	9982437658213039871725064756920320000
11	776966836171770144107444346734230682311065600000

Can we find estimates for $L(n)$, perhaps upper and lower bounds which are not too far apart?

The number of ways of placing symbols into an $n \times n$ array without restriction is n^{n^2}; so clearly $L(n) \leq n^{n^2}$. We can improve this by observing that each row is a permutation of $\{1, \ldots, n\}$, so that $L(n) \leq (n!)^n$. By Stirling's formula, the right-hand side is about $(2\pi n)^{n/2} (n/e)^{n^2}$. So, roughly speaking, we have knocked off a factor of e^{n^2}.

This bound can be further improved by noting that the second, $\ldots$, nth rows are derangements of the first; so $L(n) \leq n! d(n)^{n-1}$. The improvement is only a factor of about e^{n-1}.

We can find a lower bound by building Latin squares row by row. We define a $k \times n$ *Latin rectangle* for $k \leq n$ to be a $k \times n$ array containing the symbols $\{1, \ldots, n\}$ so that each symbol occurs once in each row and at most once in each column. Note that a $1 \times n$ Latin rectangle is simply a permutation; there are $n!$ of these. A $2 \times n$ Latin rectangle consists of two permutations, the second a derangement of the first; the number of these is $n! d(n) \sim (n!)^2 / e$.

Proposition 5.6 *For $k < n$, any $k \times n$ Latin rectangle can be extended to a $(k+1) \times n$ Latin rectangle; the number of such extensions is the permanent of a suitable $n \times n$ matrix of zeros and ones with row and column sums $n - k$.*

Proof Given a $k \times n$ Latin rectangle L, let A_i be the set of symbols which do not occur in the ith column. Then the ith entry in the $(k+1)$st row must be chosen from A_i; and these entries must all be distinct, so the row is a SDR for the sets $(A_1, \ldots, A_n)$. Thus the number of ways to extend the array to one of size $(k+1) \times n$ is the number of SDRs of $(A_1, \ldots, A_n)$, which is equal to the permanent of its incidence matrix P.

We claim that P has all row and column sums $n - k$. For the rows this is clear, since $|A_i| = n - k$ for all i. The jth column sum is the number of columns in which j does not occur in the array. But j occurs k times in the array, in k distinct columns, so fails to appear in $(n - k)$ columns.

The matrix $(1/(n - k))P$ in the above proof is doubly stochastic. By the theorem of Egorychev and Falikman, $\mathrm{per}((1/(n - k))P) \geq n!/n^n$; so we have

$$\mathrm{per}(P) \geq n!\,(n - k)^n/n^n.$$

In the case $k = 1$, assuming that the first row is the identity permutation, we have $P = J - I$, where I is the matrix with every entry 1. A permutation contributes to the permanent of P if and only if it is a derangement; so $\mathrm{per}(P) = d(n) \sim n!/e$. The estimate from the Egorychev–Falikman Theorem is

$$n!\,(1 - 1/n)^n \sim n!/e.$$

since $(1 - 1/n) \to e$ as $n \to \infty$.

For $n = 10$, we have $(9/10)^{10} = 0.3486784401$, while $1/e = 0.367879441\ldots$; so our estimate from van der Waerden's conjecture is about 5% too low. The approximation gets better for larger n.

Now the total number of Latin squares satisfies

$$L(n) \geq \prod_{k=0}^{n-1} \frac{n!\,(n - k)^n}{n^n} = \frac{(n!)^{2n}}{n^{n^2}}.$$

The number on the right is about $(2\pi n)^n (n/e^2)^{n^2}$, by Stirling's formula.

We have obtained upper and lower bounds for $L(n)$ which are roughly $(cn)^{n^2}$, where $c = 1/e^2$ for the lower bound and $1/e$ for the upper. Existing exact results are not adequate for making guesses about which bound is closer to the truth!

Remark For $k \geq 2$, the number of extensions of a $k \times n$ Latin rectangle to a $(k + 1) \times n$ Latin rectangle is not constant. For example, the two Latin rectangles

1	2	3	4
2	3	4	1

and

1	2	3	4
2	1	4	3

have 2 and 4 extensions respectively. However, it is conjectured that the number of extensions of a $2 \times n$ Latin rectangle to a Latin square is approximately constant, in the sense that the ratio of the largest to the smallest tends (rapidly) to 1 as $n \to \infty$.

5.5 Latin rectangles

The number of $1 \times n$ Latin rectangles is $n!$ (these are simply the permutations), and the number of $2 \times n$ Latin rectangles is $n! \, d(n)$, where $d(n)$ is the derangement number.

It is more difficult to count $3 \times n$ rectangles; some special types have been enumerated. Consider the case where the second row is a cyclic shift of the first. We may assume that the first row is $(1, 2, \ldots, n)$ (to get the total, we simply multiply the number of rectangles with this first row by $n!$), and the second row $(2, 3, \ldots, n, 1)$. We need to count the number of ways of adding the third row.

This problem is known as the *problème des ménages*, and can be formulated as follows. A number n of couples are to be seated at $2n$ places round a circular table. The gentlemen take alternate seats. In how many ways can the ladies be seated so that no lady is seated next to her partner? If we number the couples from 1 to n so that the gentlemen are seated in the order $(1, \ldots, n)$ around the table, then the lady a_i who sits in the place between gentlemen numbers i and $i + 1$ should not be the partner of either of these; that is, $a_i \notin \{i, i+1\}$ for $i = 1, \ldots, n$ (where the indices are taken modulo n).

We will solve this counting problem in Chapter 8, using the Inclusion–Exclusion Principle.

5.6 Exercises

Let $d(n)$ denote the number of derangements of $\{1, \ldots, n\}$.

5.1 Show that the permanent of the $n \times n$ matrix with diagonal entries 0 and off-diagonal entries 1 is equal to $d(n)$.

5.2 Show that the number of $2 \times n$ Latin rectangles is equal to $n! \, d(n)$.

5.3 Show that it is not possible to construct Latin squares by placing elements from $\{1, \ldots, n\}$ into an $n \times n$ grid arbitrarily subject to the constraint that no element occurs more than once in the same row or column. What is the smallest number of elements which have to be added in order to reach a situation where completion to a Latin square is impossible?

q-analogues

Much of the enumerative combinatorics of sets and functions can be generalised in a manner which, at first sight, seems a bit unmotivated. We introduce a new parameter q and construct functions which, in the limit as q tends to 1, become familiar combinatorial numbers such as factorials and binomial coefficients.

In this chapter, we develop a small amount of this large body of theory.

6.1   Motivation

We can look at q-analogues in several ways:

- The q-analogues are, typically, formulae which tend to the classical ones as $q \to 1$. Most basic is the fact that

$$\lim_{q \to 1} \frac{q^a - 1}{q - 1} = a$$

  for any real number a (this is immediate from l'Hôpital's rule).

- There is a formal similarity between statements about subsets of a set and subspaces of a vector space, with cardinality replaced by dimension. For example, the Inclusion–Exclusion rule

$$|U \cup V| + |U \cap V| = |U| + |V|$$

  for sets becomes

$$\dim(U + V) + \dim(U \cap V) = \dim(U) + \dim(V)$$

  for vector spaces. Now, if the underlying field has q elements, then the number of 1-dimensional subspaces of an n-dimensional vector space is $(q^n - 1)/(q - 1)$, which is exactly the q-analogue of n.

- Sometimes, replacing a simple count of objects with summing the weights of objects weighted in a certain way can translate binomial coefficients into Gaussian coefficients. We see an example involving lattice paths shortly.

- There is a notion of 'quantum calculus' or 'q-calculus', which is described briefly in the last section of this chapter.

- The simplest example of non-commutative geometry involves an algebra generated by two indeterminates x and y satisfying $yx = qxy$. In such an algebra, we can rearrange any monomial of degree n in x and y to the form $x^{n-i}y^i$, and so there will be an expression

$$(x+y)^n = \sum_{i=0}^n c_q(n,i)x^{n-i}y^i.$$

The coefficients $c_q(n,i)$ will be q-analogues of binomial coefficients.

- There are applications in algebra, where for example it is easier to find the irreducible representations of an algebra depending on a parameter q and study them as $q \to 1$ than to find the representations of the limit algebra directly.

- The analogy can be interpreted at a much higher level, in the language of *braided categories*. I will not pursue this here. You can read more in various papers of Shahn Majid, for example Braided Groups, *J. Pure Appl. Algebra* **86** (1993), 187–221; Free braided differential calculus, braided binomial theorem and the braided exponential map, *J. Math. Phys.* **34** (1993), 4843–4856.

In connection with the second interpretation, note the theorem of Galois:

Theorem 6.1 *The cardinality of any finite field is a prime power. Moreover, for any prime power q, there is a unique field with q elements, up to isomorphism.*

To commemorate Galois, finite fields are called *Galois fields*, and the field with q elements is denoted by $\mathrm{GF}(q)$.

Here is an example where Gaussian coefficients arise in a weighted counting problem.

A *lattice path* in the first quadrant of the plane is a path starting at the origin, each step being either one unit to the right, or one unit upwards. The number of lattice paths from the origin to the point (m,n) is $\binom{m+n}{m}$. For to move from the origin to (m,n), we must take $m+n$ steps, and exactly m of these steps must be to the right; we can choose which m of the $m+n$ steps are to the right arbitrarily.

This fact can also be proved by means of a recurrence relation. Let $P(m,n)$ be the number of lattice paths from the origin to (m,n). Clearly $P(0,n) = P(m,0) = 0$.

If $m, n > 0$, then any path to (m, n) must pass either through $(m-1, n)$ or through $(m, n-1)$; so we have

$$P(m, n) = P(m-1, n) + P(m, n-1) \qquad \text{for} \qquad m, n > 0.$$

This recurrence relation determines $P(m, n)$ for all $m, n \geq 0$; and an easy induction on $m + n$ shows that $P(m, n) = \binom{m+n}{m}$.

Now suppose that we give a 'weight' of q^A to a lattice path if the area under the path is A (see Figure 6.1). What is the sum of the weights of the lattice paths? Let $Q(m, n)$ be this number. Again we have $Q(m, 0) = Q(0, n) = 1$. If $m, n > 0$, then the sum of the weights of lattice paths from the origin to $(m-1, n)$ is $Q(m-1, n)$, and the last step to (m, n) adds n to the area, and so multiplies the sum of weights by q^n. On the other hand, the sum of the weights of the paths to $(m, n-1)$ is $Q(m, n-1)$, and the final vertical step to (m, n) adds no extra area to such paths. So

$$Q(m, n) = q^n Q(m-1, n) + Q(m, n-1) \qquad \text{for} \qquad m, n > 0.$$

Again, $Q(m, n)$ is determined for all $m, n \geq 0$, and the same induction (using Proposition 6.3 in place of the binomial recurrence) shows that $Q(m, n) = \begin{bmatrix} m+n \\ m \end{bmatrix}_q$ for $m, n \geq 0$.

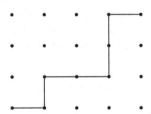

Figure 6.1: A lattice path of weight q^5

6.2   *q-integers*

We define the *q-integer* $[n]_q$ to be

$$[n]_q = \frac{q^n - 1}{q - 1}.$$

We make two observations:

(a) We have $\lim_{q \to 1} [n]_q = n$, by l'Hôpital's rule (as noted above).

(b) $[n]_q$ is the number of 1-dimensional subspaces of the *n*-dimensional vector space over $GF(q)$, for a prime power q. For such a vector space is isomorphic to the space $GF(q)^n$ of all *n*-tuples over $GF(q)$, and so has cardinality q^n. Any non-zero vector spans a 1-dimensional subspace, and any 1-dimensional subspace contains $q - 1$ non-zero vectors, each of which spans it.

In terms of *q*-integers, we can define *q*-analogues of other combinatorial functions. For example, we define the *q-factorial* to be

$$[n]!_q = [n]_q \cdot [n-1]_q \cdot [1]_q,$$

and the *q-binomial coefficient* or *Gaussian coefficient* to be

$$\begin{bmatrix} n \\ k \end{bmatrix}_q = \frac{[n]!_q}{[k]!_q [n-k]!_q}.$$

These have the property that they tend to the usual factorials and binomial coefficients as $q \to 1$. However, they will not always be the most appropriate definitions. Indeed, it is not clear that there should always be a most appropriate definition. We should proceed with caution!

It can be shown that the Gaussian coefficient is a polynomial in q, if we regard q as an indeterminate. If instead we regard q as a complex number, it has a well-defined value as long as q is not a *d*th root of unity for some d dividing k. (In the excluded cases, the denominator is zero, but the limit still exists.)

Fortunately, in the case of *q*-binomial coefficients, the proposed definition works without problems:

Proposition 6.2 *If q is a prime power, then the number of k-dimensional subspaces of an n-dimensional vector space over* $GF(q)$ *is* $\begin{bmatrix} n \\ k \end{bmatrix}_q$.

Proof The number of choices of k linearly independent vectors in $GF(q)^n$ is

$$(q^n - 1)(q^n - q) \cdots (q^n - q^{k-1}),$$

since the *i*th vector must be chosen outside the span of its predecessors. Any such choice is the basis of a unique *k*-dimensional subspace. Putting $n = k$, we see that the number of bases of a *k*-dimensional space is

$$(q^k - 1)(q^k - q) \cdots (q^k - q^{k-1}).$$

Dividing and cancelling powers of q gives the result.

6.3   The *q*-Binomial Theorem

The *q*-binomial coefficients satisfy an analogue of the recurrence relation for binomial coefficients.

Proposition 6.3 $\begin{bmatrix} n \\ 0 \end{bmatrix}_q = \begin{bmatrix} n \\ n \end{bmatrix}_q = 1, \quad \begin{bmatrix} n \\ k \end{bmatrix}_q = \begin{bmatrix} n-1 \\ k-1 \end{bmatrix}_q + q^k \begin{bmatrix} n-1 \\ k \end{bmatrix}_q \quad for\ 0 < k < n.$

Proof   This comes straight from the definition. Suppose that $0 < k < n$. Then

$$\begin{bmatrix} n \\ k \end{bmatrix}_q - \begin{bmatrix} n-1 \\ k-1 \end{bmatrix}_q = \left(\frac{q^n - 1}{q^k - 1} - 1 \right) \begin{bmatrix} n-1 \\ k-1 \end{bmatrix}_q$$

$$= q^k \left(\frac{q^{n-k} - 1}{q^k - 1} \right) \begin{bmatrix} n-1 \\ k-1 \end{bmatrix}_q$$

$$= q^k \begin{bmatrix} n \\ k-1 \end{bmatrix}_q.$$

The array of Gaussian coefficients has the same symmetry as that of binomial coefficients. From this we can deduce another recurrence relation.

Proposition 6.4   *(a) For $0 \le k \le n$,*

$$\begin{bmatrix} n \\ k \end{bmatrix}_q = \begin{bmatrix} n \\ n-k \end{bmatrix}_q.$$

(b) For $0 < k < n$,

$$\begin{bmatrix} n \\ k \end{bmatrix}_q = q^{n-k} \begin{bmatrix} n-1 \\ k-1 \end{bmatrix}_q + \begin{bmatrix} n-1 \\ k \end{bmatrix}_q.$$

Proof   (a) is immediate from the definition. For (b),

$$\begin{bmatrix} n \\ k \end{bmatrix}_q = \begin{bmatrix} n \\ n-k \end{bmatrix}_q$$

$$= \begin{bmatrix} n-1 \\ n-k-1 \end{bmatrix}_q + q^{n-k} \begin{bmatrix} n-1 \\ n-k \end{bmatrix}_q$$

$$= \begin{bmatrix} n-1 \\ k \end{bmatrix}_q + q^{n-k} \begin{bmatrix} n-1 \\ k-1 \end{bmatrix}_q.$$

We come now to the *q*-analogue of the Binomial Theorem, which states the following.

Theorem 6.5 *For a positive integer n, a real number q $\neq$ 1, and an indeterminate z, we have*

$$\prod_{i=1}^{n}(1+q^{i-1}z) = \sum_{k=0}^{n} q^{k(k-1)/2} z^k \begin{bmatrix} n \\ k \end{bmatrix}_q.$$

Proof The proof is by induction on n; starting the induction at $n = 1$ is trivial. Suppose that the result is true for $n - 1$. For the inductive step, we must compute

$$\left(\sum_{k=0}^{n-1} q^{k(k-1)/2} z^k \begin{bmatrix} n-1 \\ k \end{bmatrix}_q\right)(1+q^{n-1}z).$$

The coefficient of z^k in this expression is

$$q^{k(k-1)/2} \begin{bmatrix} n-1 \\ k \end{bmatrix}_q + q^{(k-1)(k-2)/2+n-1} \begin{bmatrix} n-1 \\ k-1 \end{bmatrix}_q$$

$$= q^{k(k-1)/2} \left(\begin{bmatrix} n-1 \\ k \end{bmatrix}_q + q^{n-k} \begin{bmatrix} n-1 \\ k-1 \end{bmatrix}_q\right)$$

$$= q^{k(k-1)/2} \begin{bmatrix} n \\ k \end{bmatrix}_q$$

by Proposition 6.4(b).

Remark The usual binomial formula appears unexpectedly in a different, though related, context, that of non-commutative geometry mentioned earlier. Suppose that, instead of being real or complex numbers, x and y are indeterminates (elements of some real algebra) which satisfy the commutation relation

$$yx = qxy.$$

Then

$$(x+y)^n = \sum_{k=0}^{n} \begin{bmatrix} n \\ k \end{bmatrix}_q x^k y^{n-k}.$$

(See Exercise 6.9.)

6.4 Elementary symmetric functions

In this section we touch briefly on the theory of elementary symmetric functions.

Let $x_1, \ldots, x_n$ be n indeterminates. For $1 \leq k \leq n$, the kth *elementary symmetric function* $e_k(x_1, \ldots, x_n)$ is the sum of all monomials which can be formed by multiplying together k *distinct* indeterminates. Thus, e_k has $\binom{n}{k}$ terms, and

$$e_k(1, 1, \ldots, 1) = \binom{n}{k}.$$

For example, if $n = 3$, the elementary symmetric functions are

$$e_1 = x_1 + x_2 + x_3, \quad e_2 = x_1 x_2 + x_2 x_3 + x_3 x_1, \quad e_3 = x_1 x_2 x_3.$$

We adopt the convention that $e_0 = 1$.

Newton observed that the coefficients of a polynomial of degree n are the elementary symmetric functions of its roots, with appropriate signs:

Proposition 6.6 $\displaystyle \prod_{i=1}^{n}(z - x_i) = \sum_{k=0}^{n}(-1)^k e_k(x_1, \ldots, x_n) z^{n-k}.$

Consider the generating function for the e_k:

$$E(z) = \sum_{k=0}^{n} e_k(x_1, \ldots, x_n) z^k.$$

A slight rewriting of Newton's Theorem shows that

$$E(z) = \prod_{i=1}^{n}(1 + x_i z).$$

Hence the Binomial Theorem and its q-analogue give the following specialisations:

Proposition 6.7   (a) *If* $x_1 = \ldots = x_n = 1$, *then*

$$E(z) = (1 + z)^n = \sum_{k=0}^{n} \binom{n}{k} z^k,$$

so

$$e_k(1, 1, \ldots, 1) = \binom{n}{k}.$$

(b) *If* $x_i = q^{i-1}$ *for* $i = 1, \ldots, n$, *then*

$$E(z) = \prod_{i=1}^{n}(1 + q^{i-1} z) = \sum_{k=0}^{n} q^{k(k-1)/2} z^k \begin{bmatrix} n \\ k \end{bmatrix}_q,$$

so

$$e_k(1, q, \ldots, q^{n-1}) = q^{k(k-1)/2} \begin{bmatrix} n \\ k \end{bmatrix}_q.$$

6.5 Partitions and permutations

The number of permutations of an n-set is $n!$. The linear analogue of this is the number of linear isomorphisms from an n-dimensional vector space to itself; this is equal to the number of choices of basis for the n-dimensional space, which is

$$(q^n - 1)(q^n - q) \cdots (q^n - q^{n-1}).$$

These linear maps form a group, the *general linear group* GL(n, q).

To an algebraist, the general linear group might seem to be a q-analogue of the symmetric group, but its order is not the q-factorial defined earlier!

Using the q-Binomial Theorem, we can transform the multiplicative formula for the order of GL(n, q) into an additive formula:

Proposition 6.8

$$|\text{GL}(n,q)| = (-1)^n q^{n(n-1)/2} \sum_{i=0}^{n} (-1)^k q^{k(k+1)/2} \begin{bmatrix} n \\ k \end{bmatrix}_q.$$

Proof We have

$$|\text{GL}(n,q)| = (-1)^n q^{n(n-1)/2} \prod_{i=1}^{n} (1 - q^i),$$

and the right-hand side is obtained by substituting $z = -q$ in the q-Binomial Theorem.

The total number of $n \times n$ matrices is q^{n^2}, so the probability that a random matrix is invertible is

$$p_n(q) = \prod_{i=1}^{n} (1 - q^{-i}).$$

As $n \to \infty$, we have

$$p_n(q) \to p(q) = \prod_{i \geq 1} (1 - q^{-i}). \tag{6.1}$$

According to Euler's Pentagonal Numbers Theorem (Proposition 4.11), we have

$$p(q) = \sum_{k \in \mathbb{Z}} (-1)^k q^{-k(3k-1)/2} = 1 - q^{-1} - q^{-2} + q^{-5} + q^{-7} - q^{-12} - \cdots \tag{6.2}$$

So, for example, $p(2) = 0.2887\ldots$ is the limiting probability that a large random matrix over GF(2) is invertible.

What is the q-analogue of the Stirling number $S(n,k)$, the number of partitions of an n-set into k parts? This is a philosophical, not a mathematical question; I argue that the q-analogue is the Gaussian coefficient $\begin{bmatrix} n \\ k \end{bmatrix}_q$.

The number of surjective maps from an n-set to a k-set is $k!S(n,k)$, since the preimages of the points in the k-set form a partition of the n-set whose k parts can be mapped to the k-set in any order. The q-analogue is the number of surjective linear maps from an n-space V to a k-space W. Such a map is determined by its kernel U, an $(n-k)$-dimensional subspace of V, and a linear isomorphism from V/U to W. So the analogue of $S(n,k)$ is the number of choices of U, which is

$$\begin{bmatrix} n \\ n-k \end{bmatrix}_q = \begin{bmatrix} n \\ k \end{bmatrix}_q.$$

6.6   Irreducible polynomials

Though it is not really a q-analogue of a classical result, the following theorem comes up in various places. Recall that a polynomial of degree n is *monic* if the coefficient of x^n is equal to 1.

Theorem 6.9 *The number $f_q(n)$ of monic irreducible polynomials of degree n over* GF(q) *satisfies*

$$\sum_{k|n} k f_q(k) = q^n.$$

Proof   We give two proofs, one depending on some algebra, and the other a rather nice exercise in manipulating formal power series.

First proof: We use the fact that the roots of an irreducible polynomial of degree k over GF(q) lie in the unique field GF(q^k) of degree k over GF(q). Moreover, GF(q^k) $\subseteq$ GF(q^n) if and only if $k \mid n$; and every element of GF(q^n) generates some subfield over GF(q), which has the form GF(q^k) for some k dividing n.

Now each of the q^n elements of GF(q^n) satisfies a unique minimal polynomial of degree k for some k; and every irreducible polynomial arises in this way, and has k distinct roots. So the result holds.

Second proof: All the algebra we use in this proof is that each monic polynomial of degree n can be factorised uniquely into monic irreducible factors. If the number of monic irreducibles of degree k is m_k, then we obtain all monic polynomials of degree n by the following procedure:

- Express $n = \sum a_k k$, where a_k are non-negative integers;

- Choose a_k monic irreducibles of degree k from the set of all m_k such, with repetitions allowed and order not important;

- Multiply the chosen polynomials together.

Altogether there are q^n monic polynomials $x^n + c_1 x^{n-1} + \cdots + c_n$ of degree n, since there are q choices for each of the n coefficients. Hence

$$q^n = \sum \prod_k \binom{m_k + a_k - 1}{a_k}, \tag{6.3}$$

where the sum is over all sequences $a_1, a_2, \ldots$ of natural numbers which satisfy $\sum k a_k = n$.

Multiplying by x^n and summing over n, we get

$$
\begin{aligned}
\frac{1}{1 - qx} &= \sum_{n \geq 0} q^n x^n \\
&= \sum_{a_1, a_2, \ldots} \prod_{k \geq 1} \binom{m_k + a_k - 1}{a_k} x^{k a_k} \\
&= \prod_{k \geq 1} \sum_{a \geq 0} \binom{m_k + a - 1}{a} (x^k)^a \\
&= \prod_{k \geq 1} (1 - x^k)^{-m_k}.
\end{aligned}
$$

Here the manipulations are similar to those for the sum of cycle indices which we will meet in the next chapter; we use the fact that the number of choices of a things from a set of m, with repetition allowed and order unimportant, is

$$\binom{m + a - 1}{a} = (-1)^a \binom{-m}{a}$$

(see Theorem 3.6), and in the fourth line we invoke the Binomial Theorem with negative exponent.

Taking logarithms of both sides, we obtain

$$
\begin{aligned}
\sum_{n \geq 1} \frac{q^n x^n}{n} &= -\log(1 - qx) \\
&= \sum_{k \geq 1} -m_k \log(1 - x^k) \\
&= \sum_{k \geq 1} m_k \sum_{r \geq 1} \frac{x^{kr}}{r}.
\end{aligned}
$$

The coefficient of x^n in the last expression is the sum, over all divisors k of n, of $m_k / r = k m_k / n$. This must be equal to the coefficient on the left, which is q^n / n.

We conclude that

$$q^n = \sum_{k|n} k m_k, \qquad (6.4)$$

as required.

Note how the very complicated recurrence relation (6.3) for the numbers m_k changes into the much simpler recurrence relation (6.4) after taking logarithms! We will see how to solve such a recurrence in the chapter on Möbius inversion.

6.7   Quantum calculus

The *q-analogues* of combinatorial formulae arise naturally in the so-called *q-calculus*, which we now describe. For further details and a wealth of examples from classical mathematics, we refer to the book *Quantum Calculus* by Victor Kac and Pokman Cheung, listed in the bibliography.

Given an arbitrary real function $f(x)$, and a real number $q \neq 1$, we define the *q-differential* to be

$$d_q f(x) = f(qx) - f(x).$$

Clearly we have $d_q x = (q-1)x$. Then we define the *q-derivative* of $f(x)$ by

$$D_q f(x) = \frac{d_q f(x)}{d_q x}.$$

Note that, unlike ordinary calculus, the differentials are defined directly, and the derivative is their quotient. If $f(x)$ is differentiable, with derivative $f'(x)$, then we have

$$\lim_{q \to 1} D_q f(x) = f'(x).$$

Now it is easy to see that

$$D_q x^n = [n]_q x^{n-1}$$

for positive integers n, where

$$[n]_q = \frac{q^n - 1}{q - 1}$$

is the *q-analogue* of n.

The *chain rule* does not hold for the *q*-calculus; it is not true that $D_q (x + a)^n = [n]_q (x+a)^{n-1}$. We need a *q*-analogue of $(x+a)^n$, which turns out to be the following function:

$$(x+a)_q^n = (x+a)(x+qa)\cdots(x+q^{n-1}a).$$

Then we do indeed have

$$D_q \, (x+a)_q^n = [n]_q \, (x+a)_q^{n-1}.$$

Now define the q-analogues of factorials and binomial coefficients by

$$[n]!_q \; = \; [1]_q \cdot [2]_q \cdots [n]_q,$$
$$\begin{bmatrix} n \\ k \end{bmatrix}_q \; = \; \frac{[n]!_q}{[k]!_q \, [n-k]!_q}.$$

Then the q-*Binomial Theorem* can be stated in the form

$$(x+y)_q^n = \sum_{k=0}^n \begin{bmatrix} n \\ k \end{bmatrix}_q q^{k(k-1)/2} x^k y^{n-k}.$$

Remark There is another type of quantum calculus, namely the *h-calculus*, defined as follows: we have

$$d_h \, f(x) = f(x+h) - f(x),$$

so that $d_h \, x = h$, and then

$$D_h \, f(x) = \frac{d_h \, f(x)}{d_h \, x}.$$

Clearly $\lim_{h \to 0} D_h \, f(x) = f'(x)$ if f is differentiable. The h-calculus is connected with the Bernoulli numbers and the Euler–Maclaurin sum formula, which we will discuss in a later chapter. For further details of these two quantum calculi and a generalisation, see the book by Kac and Cheung.

6.8   Exercises

6.1 Prove that $\begin{bmatrix} n \\ k \end{bmatrix}_q$ is a polynomial of degree $k(n-k)$ in the indeterminate q.

6.2 Prove that, for fixed n, the Gaussian coefficients $\begin{bmatrix} n \\ k \end{bmatrix}_q$ for $k = 0, 1, \ldots, n$ form a log-concave sequence.

6.3   (a) Prove that, for $0 < k < n$,

$$\begin{bmatrix} n \\ k \end{bmatrix}_q = \begin{bmatrix} n-1 \\ k-1 \end{bmatrix}_q + \begin{bmatrix} n-1 \\ k \end{bmatrix}_q + (q^{n-1} - 1) \begin{bmatrix} n-2 \\ k-1 \end{bmatrix}_q.$$

(b) Let

$$F_q(n) = \sum_{k=0}^{n} \begin{bmatrix} n \\ k \end{bmatrix}_q,$$

so that, if q is a prime power, then $F_q(n)$ is the total number of subspaces of an n-dimensional vector space over $\mathrm{GF}(q)$. Prove that

$$F_q(0) = 1, F_q(1) = 2, \quad F_q(n) = 2F_q(n-1) + (q^{n-1} - 1)F_q(n-2) \text{ for } n \geq 2.$$

(c) Deduce that, if $q > 1$, then $F_q(n) \geq c\, q^{n^2/4}$ for some constant c (depending on q).

6.4 This exercise shows that the Gaussian coefficients have a counting interpretation for all positive integer values of q (not just prime powers).

Suppose that q is an integer greater than 1. Let Q be a finite set of cardinality q containing two distinguished elements 0 and 1. We say that a $k \times n$ matrix with entries from Q is in *reduced echelon form* if the following conditions hold:

- If a row has any non-zero entries, then the first such entry is 1 (such entries are called 'leading 1');

- if $i < j$ and row j is non-zero, then row i is also non-zero, and its leading 1 occurs to the left of the leading 1 in row j;

- if a column contains the leading 1 of some row, then all other entries in that column are 0.

Prove that $\begin{bmatrix} n \\ k \end{bmatrix}_q$ is the number of $k \times n$ matrices in reduced echelon form with no rows of zeros.

6.5 A matrix is said to be in *echelon form* if it satisfies the first two conditions in the definition of reduced echelon form. Show that, if q is an integer greater than 2, the right-hand side of the q-Binomial Theorem with $x = 1$ counts the number of $n \times n$ matrices in echelon form.

How many $n \times n$ matrices in reduced echelon form are there?

6.6 Let $h_k(x_1, \ldots, x_n)$ be the *complete symmetric function* of degree k in the indeterminates $x_1, \ldots, x_n$ (the sum of *all* monomials of degree k that can be formed using these indeterminates). For example,

$$h_2(x_1, x_2, x_3) = x_1^2 + x_2^2 + x_3^2 + x_1 x_2 + x_2 x_3 + x_3 x_1.$$

Prove that

$$\sum_{k=0}^{\infty} h_k(x_1, \ldots, x_n) z^k = \prod_{i=1}^{n} (1 - x_i z)^{-1}.$$

Deduce that

(a) $h_k(1,1,\ldots,1) = \dbinom{n+k-1}{k}$;

(b) $h_k(1,q,\ldots,q^{n-1}) = \begin{bmatrix} n+k-1 \\ k \end{bmatrix}_q$ for $q \neq 1$.

Hint for (b): show that

$$\sum_{i=0}^{k} q^i \begin{bmatrix} n+i-2 \\ i \end{bmatrix}_q = \begin{bmatrix} n+k-1 \\ k \end{bmatrix}_q.$$

6.7 The second proof of Theorem 6.9 shows that the number of irreducible polynomials over $\mathrm{GF}(q)$ is exactly what is required if every element of $\mathrm{GF}(q^n)$ is the root of a unique irreducible of degree dividing n over $\mathrm{GF}(q)$. Turn the argument around to give a counting proof of the existence and uniqueness of $\mathrm{GF}(q^n)$, given that of $\mathrm{GF}(q)$.

6.8 Let ω be a primitive dth root of unity. Express $\begin{bmatrix} n \\ k \end{bmatrix}_\omega$ in terms of binomial coefficients (whenever you can).

Solution by Pablo Spiga   Let d be a natural number, and let ω be a primitive dth root of unity in $\mathbb{C}$, i.e. $\omega^d = 1$. Then, if $0 \leq a, b \leq d-1$, we have

$$\begin{bmatrix} nd+a \\ kd+b \end{bmatrix}_\omega = \binom{n}{k} \begin{bmatrix} a \\ b \end{bmatrix}_\omega.$$

Note that we are assuming that $\begin{bmatrix} a \\ b \end{bmatrix}_\omega = 0$ whenever $a < b$.

Solution   By induction on a. We have

$$\begin{aligned}
1 - \xi^d &= \prod_{i=1}^{d}(\omega^{i-1} - \xi) \\
&= \prod_{i=1}^{d}(\omega^{i-1} \cdot (1 - \omega^{-i+1}\xi)) \\
&= \prod_{i=1}^{d}\omega^{i-1} \cdot \prod_{i=1}^{d}(1 - \omega^{i-1}\xi).
\end{aligned}$$

Thus, we get

$$\prod_{i=1}^{nd}(1 + \omega^{i-1}(-\xi)) = \sum_{j=0}^{nd} \omega^{j(j-1)/2}(-1)^j \begin{bmatrix} nd \\ j \end{bmatrix}_\omega \xi^j, \tag{6.5}$$

but

$$\prod_{i=1}^{nd}(1 + \omega^{i-1}(-\xi)) = (1 - \xi^d)^n = \sum_{k=0}^{n}\binom{n}{k}(-1)^k\xi^{kd}. \tag{6.6}$$

We have proved that $\left[{nd \atop j}\right]_\omega = 0$ if d does not divide j. Assume $j = dk$. By (6.5) and (6.6), as

$$\omega^{dk(dk-1)/2}(-1)^{k(d+1)} = 1, \tag{6.7}$$

we get

$$\left[{nd \atop kd}\right]_\omega = \binom{n}{k}.$$

(For (6.7), note that if d is odd then $(-1)^{d+1} = 1$, while if d is even then we can write -1 as $\omega^{d/2}$, and we find $\omega^{dk(dk+d)/2} = \omega^{d^2k(k+1)/2}$.) This proves the result for $a = 0$.

Assume $a \geq 1$. If $b \neq 0$ then, by induction hypothesis and by the usual recurrence relation, we get

$$\begin{aligned}
\left[{nd+a \atop kd+b}\right]_\omega &= \left[{nd+a-1 \atop kd+b-1}\right]_\omega + \omega^{kd+b}\left[{nd+a-1 \atop kd+b}\right]_\omega \\
&= \binom{n}{k}\left[{a-1 \atop b-1}\right]_\omega + \omega^b\binom{n}{k}\left[{a-1 \atop b}\right]_\omega \\
&= \binom{n}{k}\left[{a \atop b}\right]_\omega.
\end{aligned}$$

Finally, if $b = 0$, then, as $a - 1 < d - 1$,

$$\begin{aligned}
\left[{nd+a \atop kd}\right]_\omega &= \left[{nd+a-1 \atop (k-1)d+d-1}\right]_\omega + \omega^{kd}\left[{nd+a-1 \atop kd}\right]_\omega \\
&= \binom{n}{k}\omega^0\left[{a-1 \atop 0}\right]_\omega \\
&= \binom{n}{k}\left[{a \atop b}\right]_\omega.
\end{aligned}$$

Remark Compare Lucas' formula

$$\binom{np+a}{kp+b} \equiv \binom{n}{k}\binom{a}{b} \quad (\text{mod } p)$$

if p is prime and $0 \leq a, b < p$.

6.9 Prove that, if the indetermintes x and y satisfy $yx = qxy$, then

$$(x+y)^n = \sum_{k=0}^{n}\left[{n \atop k}\right]_q x^{n-k}y^k.$$

6.10 (a) Prove that

$$D_q(f(x)g(x)) = f(qx)D_q(g(x)) + g(x)D_q(f(x)).$$

(b) Find a formula for $D_q(f(x)/g(x))$.

(c) Prove that, if $g(x) = ax^n$, then

$$D_q(f(g(x))) = (D_{q^n}f)(g(x)) \cdot D_q(g(x)).$$

6.11 Prove *Heine's formula*, the q-analogue of the *negative Binomial Theorem*:

$$\frac{1}{(1-x)_q^n} = \sum_{j=0}^{\infty} \begin{bmatrix} n+j-1 \\ j \end{bmatrix}_q x^j.$$

6.12 Prove *Euler's formulae*

$$\prod_{i=0}^{\infty}(1+xq^i) = \sum_{j=0}^{\infty} q^{j(j-1)/2} \frac{x^j}{(1-q)(1-q^2)\cdots(1-q^j)},$$

$$\prod_{i=0}^{\infty}(1-xq^i)^{-1} = \sum_{j=0}^{\infty} \frac{x^j}{(1-q)(1-q^2)\cdots(1-q^j)},$$

Hint: Let $n \to \infty$ in the q-Binomial Theorems.

6.13 Define the q-exponential by

$$\exp_q(x) = \sum_{n=0}^{\infty} \frac{x^n}{[n]!_q}.$$

Prove that, if $yx = qxy$, then $\exp_q(x+y) = \exp_q(x)\exp_q(y)$.

6.14 Using each of the formulae (6.1) and (6.2), estimate the probability that a large matrix over $GF(2)$ is invertible. Compare the rate of convergence of the two formulae.

6.15 (a) Let W be a k-dimensional subspace of an n-dimensional vector space over $GF(q)$. Prove that the number of complements of W in V is $q^{k(n-k)}$.

(b) Hence show that the number of linear bijections between pairs of k-dimensional subspaces of V is equal to the number of linear maps from V to V of rank k.

(c) Deduce that the total number of linear bijections between subspaces of V is equal to the total number of linear maps from V to V.

Remark   The result of this exercise can be expressed as the identity

$$\sum_{k=0}^{n} {\begin{bmatrix} n \\ k \end{bmatrix}}_q^2 |\text{GL}(k,q)| = q^{n^2}.$$

This is a true analogue of a 'false theorem' about sets and functions. The number of partial permutations of an n-set is

$$\sum_{k=0}^{n} {\binom{n}{k}}^2 |S_k|$$

(see Theorem 4.20), whereas the number of maps from an n-set to itself is n^n; these numbers are definitely not equal!

CHAPTER 7

Group actions and the cycle index

There are many situations in combinatorics where we want to count the number of arrangements of some kind, not in total, but up to some symmetry of the problem concerned. The branch of mathematics which deals with symmetry is group theory, in particular the theory of group actions. In this chapter we consider how to use information about a group action (specifically, a polynomial known as the *cycle index*) to solve problems of this kind.

Here is an example.

A cube has six faces, so if we paint each face red, white or blue, the total numbers of ways that we can apply the colours is $3^6 = 729$. However, if we can pick up the cube and move it around, it is natural to count in a different way, where two coloured cubes differing only by a rotation are counted as 'the same'. There are 24 rotations of the cube into itself, but the answer to our question is not obtained just by dividing 729 by 24. The purpose of this section is to develop tools for answering such questions.

The theory of permutation groups has many connections to combinatorics other than its use in enumeration. To choose one example, a number of the 'sporadic' finite simple groups were first constructed as groups of automorphisms of combinatorial structures such as graphs or designs.

7.1 Group actions

Let X be a set, and G a set of permutations of X. We write the image of $x \in X$ under the permutation g as x^g. We denote the identity permutation (leaving every element of X where it is) by 1, and the inverse of a permutation g (the permutation h with $x^g = y \Leftrightarrow x^h = x$) by g^{-1}. The composition of two permutations g and h, denoted by gh, is defined by the rule that

$$x^{gh} = (x^g)^h$$

(in other words, apply first g, then h).

We say that G is a *permutation group* if the following conditions hold:

- G contains the identity permutation;
- G contains the inverse of each of its elements;
- G contains the composition of any two of its elements.

For example, the 24 rotational symmetries of a cube form a permutation group on the set of points of the cube.

Until the middle of the nineteenth century, what we have just defined would have simply been called a *group*. Now the definition of a group is more abstract. We don't go into abstract group theory here, but note some terminology arising from this. If G is an abstract group in the modern sense, an *action* of G on the set X is a function associating a permutation of X with each group element, in such a way that the identity, inverse, and composition of permutations correspond to the same concepts in the abstract group.

In particular, if G is a permutation group on a set X, then we can construct actions of G on various auxiliary sets built from X: for example, the set of ordered pairs of elements of X, the set of subsets of X, the set of functions from X to another set (or from another set to X).

For example, G acts on the set $X \times X$ of ordered pairs of elements of X by the rule

$$(x,y)^g = (x^g, y^g)$$

for $x, y \in X$, $g \in G$; that is, the permutation g acts coordinate-wise on ordered pairs, mapping (x,y) to (x^g, y^g).

Thus, the phrases 'G is a permutation group on X' and 'G acts on X' are almost synonymous; the difference is of less interest to a combinatorialist than to an algebraist.

Suppose that G acts on X. We define a relation $\sim$ on X by the rule that $x \sim y$ if $y = x^g$ for some $g \in G$.

Proposition 7.1 $\sim$ *is an equivalence relation.*

Proof We check the three conditions.

- $x = x^1$, so $x \sim x$: $\sim$ is reflexive.
- Let $x \sim y$. Then $y = x^g$, so $x = y^{g^{-1}}$, so $y \sim x$: $\sim$ is symmetric.

- Let $x \sim y$ and $y \sim z$. Then $x = x^g$ and $z = y^h$, for some $g, h \in G$. Thus, $z = (x^g)^h = x^{gh}$, so $x \sim z$: $\sim$ is transitive.

Note that the three conditions in the definition of a permutation group translate precisely into the three conditions of an equivalence relation.

The equivalence classes of this relation are the *orbits* of G on X.

In our coloured cube example, the group of 24 rotations of the cube acts on the set of 729 colourings of the faces of the cube. Two colourings count 'the same' if and only if they are in the same orbit. So our task is to count orbits.

7.2 The Orbit-counting Lemma

For any permutation g of X, we let fix(g) denote the number of *fixed points* of g (elements $x \in X$ such that $x^g = x$).

Theorem 7.2 *(Orbit-counting Lemma) Let G be a permutation group on the finite set X. Then the number of orbits of G on X is given by the formula*

$$\frac{1}{|G|} \sum_{g \in G} \text{fix}(g).$$

Remark In the literature, this result is often referred to as *Burnside's Lemma*; see Neumann's paper in the bibliography to see why this name came about and why it is inappropriate.

Proof We count in two different ways the number N of pairs (x, g), with $x \in X$, $g \in G$, and $x^g = x$.

On the one hand, clearly

$$N = \sum_{g \in G} \text{fix}(g).$$

On the other hand, we claim that if the point x lies in an orbit $\{x = x_1, \ldots, x_n\}$, then the number of permutations $g \in G$ with $x^g = x$ is $|G|/n$. More generally, for any i with $1 \leq i \leq n$, the number of permutations $g \in G$ with $x^g = x_i$ is independent of i (the proof is an exercise), and so is $|G|/n$.

Hence the number of pairs (y, g) with $y^g = y$ for which y lies in a fixed orbit of size n is $n \cdot |G|/n = |G|$. So each orbit contributes $|G|$ to the sum, and so $N = |G|k$, where k is the number of orbits.

Equating the two values gives the result.

Using this, we can count our coloured cubes. We have to examine the 24 rotations and find the number of colourings fixed by each.

- The identity fixes all $3^6 = 729$ colourings.
- There are three axes of rotation through the mid-points of opposite faces. A rotation through a half-turn about such an axis fixes $3^4 = 81$ colourings: we can choose arbitrarily the colour for the top face, the bottom face, the east and west faces, and the north and south faces (assuming that the axis is vertical). A rotation about a quarter turn fixes $3^3 = 27$ colourings, since all four faces except top and bottom must have the same colour. There are three half-turns and six quarter-turns.
- A half-turn about the axis joining the midpoints of opposite edges fixes $3^3 = 27$ colourings. There are six such rotations.
- A third-turn about the axis joining opposite vertices fixes $3^2 = 9$ colourings. There are eight such rotations.

By Theorem 7.2, the number of orbits is

$$\frac{1}{24}(1 \cdot 729 + 3 \cdot 81 + 6 \cdot 27 + 6 \cdot 27 + 8 \cdot 9) = 57,$$

so there are 57 different colourings up to rotation.

At this point, we can give a more combinatorial proof of the formula

$$x(x-1)\cdots(x-n+1) = \sum_{k=1}^{n} s(n,k)x^k$$

from Proposition 3.18. We prove the equivalent form

$$x(x+1)\cdots(x+n-1) = \sum_{k=1}^{n} u(n,k)x^k$$

for the unsigned Stirling numbers of the first kind, from which the required equation is obtained by substituting $-x$ for x and multiplying by $(-1)^n$. Suppose first that x is a positive integer. Consider the set of functions from $\{1,\ldots,n\}$ to a set X of cardinality x. There are x^n such functions. Now the symmetric group S_n acts on these functions: the permutation g maps the function f to f^g, where

$$f^g(i) = f(ig^{-1}).$$

A function f selects n elements of X with order important and repetitions allowed: the selections are just $f(1),\ldots,f(n)$. So the orbit of f under S_n is simply a selection of n things from X, where repetitions are allowed and order is *not* important (since S_n allows all reorderings). So the number of orbits is the number of such selections, which is

$$\binom{x+n-1}{n} = x(x+1)\cdots(x+n-1)/n!$$

(see Theorem 3.6).

We can also count the orbits using the Orbit-counting Lemma. Let g be a permutation in S_n having k cycles. How many functions are fixed by g? Clearly a function f is fixed if and only if it is constant on each cycle of g (for we have $f(i) = f^g(i) = f(ig^{-1}) = f(ig^{-2}) = \cdots$). The values of f on different cycles can be chosen arbitrarily. So there are x^k fixed functions. Since the number of permutations with k cycles is $u(n,k)$, the Orbit-counting Lemma shows that the number of orbits is

$$\frac{1}{n!} \sum_{k=1}^{n} u(n,k)x^k.$$

Equating the two expressions and multiplying by $n!$ gives the result.

Now the required equation holds for all positive integer values of x, and so it is a polynomial identity.

7.3 The cycle index

It is possible to develop a method for solving the coloured cubes problem which doesn't require extensive recalculation when small changes are made (such as changing the number of colours).

Suppose that we have a set F of objects called 'figures', each of which (say f) has a non-negative integer 'weight' $w(f)$ associated with it. The number of figures may be infinite, but we assume that there are only a finite number of any given weight: say a_n figures of weight n. The *figure-counting series* is the (ordinary) generating function for these numbers:

$$A(x) = \sum_{n \geq 0} a_n x^n.$$

We attach a figure to each point of a finite set X. (Equivalently, we take a function ϕ from X to the set F of figures.) The *weight* of the function ϕ is just

$$w(\phi) = \sum_{x \in X} w(\phi(x)).$$

Finally, we have a group G of permutations of X. Then G acts on the set of functions by the rule that

$$\phi^g(x) = \phi(xg^{-1}).$$

Clearly $w(\phi^g) = w(\phi)$ for any function ϕ.

We want to find the generating function for the number of functions of each possible weight, but counting two functions as 'the same' if they lie in the same orbit of G with the above action. In other words, we want to calculate the *function-counting series*

$$B(x) = \sum_{n \geq 0} b_n x^n,$$

where b_n is the number of orbits consisting of functions of weight n.

In the coloured cubes example, if we take three figures Red, White and Blue, each of weight 0, the figure-counting series is simply 3, and the function-counting series is 57. We could, say, change the weight of Red to 1, so that the figure-counting series is $2 + x$; then the function-counting series is the generating function for the numbers of colourings with $0, 1, 2, \ldots, 6$ red faces (up to rotations).

The gadget that does this job is the *cycle index* of G. Each element $g \in G$ can be decomposed into disjoint cycles; let $c_i(g)$ be the number of cycles of length i, for $i = 1, \ldots, n = |X|$. Now put

$$z(g) = s_1^{c_1(g)} s_2^{c_2(g)} \cdots s_n^{c_n(g)},$$

where $s_1, \ldots, s_n$ are indeterminates. Then the *cycle index* of G is defined to be

$$Z(G) = \frac{1}{|G|} \sum_{g \in G} z(g).$$

For example, our analysis of the rotations of the cube shows that the cycle index of this group (acting on faces) is

$$\frac{1}{24}(s_1^6 + 3s_1^2 s_2^2 + 6s_1^2 s_4 + 6s_2^3 + 8s_3^2).$$

We use the notation

$$Z(G; s_i \leftarrow f_i \text{ for } i = 1, \ldots, n)$$

for the result of substituting the expression f_i for the indeterminate s_i for $i = 1, \ldots, n$.

Theorem 7.3 *If G acts on X, and we attach figures to the points of X with figure-counting series $A(x)$, then the function-counting series is given by*

$$B(x) = Z(G; s_i \leftarrow A(x^i) \text{ for } i = 1, \ldots, n).$$

For example, in the coloured cubes, let Red have weight 1 and the other colours weight 0. Then $A(x) = 2 + x$, and the function-counting series is

$$
\begin{aligned}
B(x) &= \frac{1}{24}((2+x)^6 + 3(2+x)^2(2+x^2)^2 + 6(2+x)^2(2+x^4) \\
&\quad + 6(2+x^2)^3 + 8(2+x^3)^2) \\
&= 10 + 12x + 16x^2 + 10x^3 + 6x^4 + 2x^5 + x^6.
\end{aligned}
$$

Note that putting $x = 1$ recovers the value 57.

Proof The first step is to note that, if we ignore the group action and simply count all the functions, the function-counting series is $B(x) = A(x)^n$, where $n = |X|$. For the term in x^m in $A(x)^n$ is obtained by taking all expressions $m = m_1 + \cdots + m_n$ for m as a sum of n non-negative integers, multiplying the corresponding terms $a_{m_i}^{m_i}$ in $A(x)$, and summing the result. The indicated product counts the number of choices of functions of weights $m_1, \ldots, m_n$ to attach at the points $1, \ldots, n$ of X, so the result is indeed the function-counting series.

Note that this proves the theorem in the case where G is the trivial group.

Next, we have to count the functions of given weight fixed by a permutation $g \in G$. As we have seen, a function is fixed by g if and only if it is constant on the cycles of g. Now if we choose a function of weight r to attach to the points of a particular i-cycle of g, the number of choices is a_r but the contribution to the weight is ir. Arguing as above, the generating function for the number of fixed functions is

$$A(x)^{c_1(g)} A(x^2)^{c_2(g)} \cdots A(x^n)^{c_n(g)} = z(g; s_i \leftarrow A(x^i) \text{ for } i = 1, \ldots, n).$$

Finally, by the Orbit-counting Lemma, if we sum over $g \in G$ and divide by $|G|$, we find that the function-counting series is

$$B(x) = Z(G; s_i \leftarrow A(x^i) \text{ for } i = 1, \ldots, n).$$

7.4 Labelled and unlabelled

Group actions can be used to clarify the difference between two types of counting of combinatorial objects, namely counting labelled and unlabelled objects.

Typically, we are counting structures 'based on' a set of n points: these may be partitions or permutations, or more elaborate relational structures such as graphs, trees, partially ordered sets, etc. An *isomorphism* between two such objects is a bijection between their base sets which preserves the structure.

A *labelled object* is simply an object whose base set is $\{1, 2, \ldots, n\}$. Two objects count as different unless they are identical. On the other hand, for unlabelled objects, we wish to count them as the same obtain one from the other by re-labelling the points of the base set. In other words, an *unlabelled object* is an isomorphism class of objects.

For example, for graphs on three vertices, there are eight labelled objects, but four unlabelled ones. Figure 7.1 shows the four unlabelled graphs; check that the numbers of different labellings (assignments of the numbers $1, 2, 3$ to the vertices) are $1, 3, 3, 1$ respectively.

Now the symmetric group S_n acts on the set of all labelled objects on the set $\{1, \ldots, n\}$; its orbits are the unlabelled objects. So counting unlabelled objects is equivalent to counting orbits of S_n in an appropriate action.

A given object A has an automorphism group $\mathrm{Aut}(A)$, consisting of all permutations of the set of points which map the object to itself. The number of different

Figure 7.1: Unlabelled graphs on 3 vertices

labellings of A is $n!/|\operatorname{Aut}(A)|$, since of the $n!$ labellings, two are the same if and only if they are related by an automorphism of A. (More formally, labellings correspond bijectively to cosets of $\operatorname{Aut}(A)$ in the symmetric group S_n.) So the number of labelled objects is

$$\sum_A \frac{n!}{|\operatorname{Aut}(A)|},$$

where the sum is over the unlabelled objects on n points.

The cycle index method can be applied to give more sophisticated counts. For example, let us count graphs on 4 vertices. The number of pairs of vertices is 6, and each pair is either an edge or a non-edge. So the number of labelled graphs is $2^6 = 64$, and the number of labelled graphs with k edges is $\binom{6}{k}$ for $k = 0, \ldots, 6$.

In order to count orbits, we must let S_4 act on the set of 64 graphs. But we can think of a graph as the set of $\binom{4}{2} = 6$ pairs of vertices with a figure (either an edge or a non-edge) attached to each. So we must compute the cycle index of S_4 acting on pairs of vertices. Table 7.1 gives details. The notation $1^2 2^1$, for example, means 'two fixed points and one 2-cycle'. Such an element, say the transposition $(1,2)$, fixes the two pairs $\{1,2\}$ and $\{3,4\}$, and permutes the other four pairs in two 2-cycles; so its cycle structure on pairs is $1^2 2^2$.

Cycles on vertices	Cycles on pairs	Number
1^4	1^6	1
$1^2 2^1$	$1^2 2^2$	6
2^2	$1^2 2^2$	3
13	3^2	8
4	24	6

Table 7.1: Cycle index of S_4

So the cycle index of the permutation group G induced on pairs by S_4 is

$$Z(G) = \frac{1}{24}(s_1^6 + 9s_1^2 s_2^2 + 8s_3^2 + 6s_2 s_4).$$

Now if we take edges to have weight 1 and non-edges to have weight 0 (that is, figure-counting series $A(x) = 1 + x$), the function-counting series is

$$B(x) = 1 + x + 2x^2 + 3x^3 + 2x^4 + x^5 + x^6,$$

the generating function for unlabelled graphs on four vertices by number of edges.

We conclude by summarising some of our earlier results on counting labelled and unlabelled structures. Table 7.2 gives the numbers of labelled and unlabelled structures on n points; $B(n)$ and $p(n)$ are the Bell and partition numbers.

Structure	Labelled	Unlabelled
Subsets	2^n	$n+1$
Partitions	$B(n)$	$p(n)$
Permutations	$n!$	$p(n)$
Total orders	$n!$	1

Table 7.2: Labelled and unlabelled

We see from the table that it is possible, even in very natural cases, to have the same number of labelled objects but different numbers of unlabelled ones, or *vice versa*.

7.5   Exercises

7.1 Use the Cycle Index Theorem to write down a polynomial in two variables x and y in which the coefficient of $x^i y^j$ is the number of cubes in which the faces are coloured red, white and blue, having i red and j blue faces, up to rotations of the cube.

7.2 Find a formula for the number of ways of colouring the faces of the cube with r colours, up to rotations of the cube. Repeat this exercise for the other four Platonic solids.

7.3 A necklace has ten beads, each of which is either black or white, arranged on a loop of string. A cyclic permutation of the beads counts as the same necklace. How many necklaces are there? How many are there if the necklace obtained by turning over the given one is regarded as the same?

7.4 Repeat this question for necklaces with n beads of r possible colours.

7.5 Let G act transitively on X, where $|X| = n > 1$.

(a) By considering the action of G on ordered pairs of elements of X, show that

$$\frac{1}{|G|} \sum_{g \in G} \text{fix}(g)^2 \geq 2.$$

(b) Hence show that

$$\frac{1}{|G|} \sum_{g \in G} (\mathrm{fix}(g) - 1)(\mathrm{fix}(g) - n) \geq 1.$$

(c) Deduce that the proportion of elements of G with no fixed points is at least $1/n$.

(This result is due to Cameron and Cohen.)

7.6 Let G be a permutation group on a set X, where $|X| = n$.

For $0 \leq i \leq n$, let p_i be the proportion of elements of G which have exactly i fixed points on X, and let $p(x) = \sum p_i x^i$ be the generating function for these numbers (the *probability generating function for fixed points*).

For $0 \leq i \leq n$, let F_i be the number of orbits of G in its action on the set of i-tuples of distinct elements of X, and let $F(x) = \sum F_i x^i / i!$ be the e.g.f. for these numbers.

Use the Orbit-counting Lemma to show that

$$F(x) = P(x+1)$$

and deduce that the proportion of fixed-point-free elements in G is $p_0 = F(-1)$.

Taking G to be the symmetric group S_n, deduce the formula for the derangement number d_n (see Section 4.2.1).

(This result is due to Boston *et al.*)

7.7 A necklace is made with two different colours of beads; there are N beads altogether. We specify that there are to be k points on the necklace where the colours change (that is, k maximal runs of beads of the same colour), where k is even. Show that, if the two colours are interchangeable, then the number of different necklaces which can be produced is the coefficient of x^N in

$$Z(G; s_i \leftarrow x^i / (1 - x^i), 1 \leq i \leq k),$$

where G is the dihedral group of order $2k$.

CHAPTER 8

Möbius inversion

Often we are in the situation where we have a number of conditions of varying strength, and we have information about the number of objects which satisfy various combinations of conditions (inclusion); we want to count the objects satisfying none of the conditions (exclusion), or perhaps satisfying some but not others. Of course, the conditions may not all be independent; this is what makes the problem hard. The Principle of Inclusion and Exclusion solves this problem. A more general form of this principle, known as *Möbius inversion*, can be used for a much more general class of problems of this type.

8.1   The Principle of Inclusion and Exclusion

Let $A_1, \ldots, A_n$ be subsets of a finite set X. For any non-empty subset J of the index set $\{1, \ldots, n\}$, we put

$$A_J = \bigcap_{j \in J} A_j;$$

by convention, we take $A_\emptyset = X$. The *Principle of Inclusion and Exclusion* (PIE, for short) asserts the following.

Theorem 8.1 *The number of elements of X lying in none of the sets A_i is equal to*

$$\sum_{J \subseteq \{1, \ldots, n\}} (-1)^{|J|} |A_J|.$$

Proof   The expression in the theorem is a linear combination of the cardinalities of the sets A_J, and so we can calculate it by working out, for each $x \in X$, the contribution of x to the sum. If K is the set of all indices j for which $x \in A_j$, then

x contributes to the terms involving sets $J \subseteq K$, and the contribution is

$$\sum_{J \subseteq K} (-1)^{|J|}.$$

If $|K| = k > 0$, then there are $\binom{k}{j}$ sets of size j in the sum, which is

$$\sum_{j=0}^{k} \binom{k}{j} (-1)^j = (1-1)^k = 0,$$

whereas if $K = \emptyset$ then the sum is 1. So the points with $K = \emptyset$ (those lying in no set A_i) each contribute 1 to the sum, and the remaining points contribute nothing. So the theorem is proved.

There is a more general form of Inclusion–Exclusion, where the elements of the set X have weights, and cardinalities of sets are replaced by the sums of the weights of their elements. We must be able to add and subtract weights, and multiply them by integers; in other words, we can take the weights from an arbitrary abelian group.

If there are numbers $m_0, \ldots, m_n$ such that $|A_J| = m_j$ whenever $|J| = j$, then PIE can be written in the simpler form

$$\sum_{j=0}^{n} (-1)^j m_j \binom{n}{j}.$$

Here are a couple of applications.

Example: Surjections The number of functions from an m-set *onto* an n-set is given by the formula

$$\sum_{j=0}^{n} (-1)^j \binom{n}{j} (n-j)^m.$$

For let M and N be the sets, with $N = \{1, \ldots, n\}$. Let X be the set of all functions $f : M \to N$, and A_i the set of functions whose range does not include the point i. Then A_J is the set of functions whose range includes none of the points of J (that is, functions from M to $N \setminus J$); so $|A_J| = (n-j)^m$ when $|J| = j$. A function is a surjection if and only if it lies in none of the sets A_i. The result follows.

In particular, if $m = n$, then surjections are permutations, and we have

$$\sum_{j=0}^{n} (-1)^j \binom{n}{j} (n-j)^n = n!.$$

Example: Derangements This time, let X be the set of all permutations of $\{1, \ldots, n\}$, and A_i the set of permutations fixing i. Then A_J is the set of permutations fixing every point in J; so $|A_J| = (n-j)!$ when $|J| = j$. The permutations lying in none of the sets A_i are the derangements, and so we have

$$d(n) = \sum_{j=0}^{n} (-1)^j \binom{n}{j} (n-j)!$$
$$= n! \sum_{j=0}^{n} \frac{(-1)^j}{j!},$$

in agreement with our earlier result.

Example: the *problème des ménages* We have some unfinished business from Chapter 5: how many possible third rows of Latin rectangles which have first row $(1, 2, \ldots, n)$ and second row $(2, 3, \ldots, n, 1)$ are there? In other words, we require a permutation $(a_1, \ldots, a_n)$ so that a_i does not belong to the ith column of the array

$$\begin{pmatrix} 1 & 2 & \ldots & n-1 & n \\ 2 & 3 & \cdots & n & 1 \end{pmatrix}.$$

We will proceed slightly differently with the count: for $r = 0, \ldots, n$, let B_r be the number of permutations $(a_1, \ldots, a_n)$ for which $a_i \in \{i, i+1\}$ holds for at least a prescribed set of r values of i.

Now $B_1 = 2n(n-1)!$; for we can choose in n ways a position i for which $a_i \in \{i, i+1\}$ holds, in two ways which value of a_i to take, and in $(n-1)!$ ways the completion of the permutation.

To compute B_r for $r > 1$, note that we may assume that either $a_1 = 1$ or $a_n = 1$. In the first case, we consider the list

$$(2, 2, 3, 3, \ldots, n-1, n-1, n)$$

consisting of the elements after the one chosen written column by column and claim that the remaining $r - 1$ positions satisfying the condition form a selection of $r - 1$ from this list of $2n - 3$ with no two consecutive. (Two consecutive numbers either correspond to two entries from the same column, or two entries with the same value.) This can be done in $\binom{2n-r-1}{r-1}$ ways (Exercise 1.1). The other choice would give the same value, using the sequence $(1, 2, 2, \ldots, n-1, n-1)$ instead.

So there are $2n$ choices of starting point and $\binom{2n-r-1}{r-1}$ ways to choose the remaining values; but we must divide by r since we arbitrarily chose one of the distinguished r points to be the start. Finally, there are $(n-r)!$ ways to choose the remaining $n - r$ entries in the permutation. So

$$B_r = \frac{2n}{r} \binom{2n-r-1}{r-1}.$$

Finally, applying PIE shows that the number of solutions of the *problème des ménages* is

$$M_n = n! - \sum_{r=1}^{n} (-1)^{r-1} \frac{2n}{r} \binom{2n-r-1}{r-1} (n-r)!.$$

The statement of PIE can be generalised to give a formula for the number of elements of X which lie in a given collection of sets A_i and not in the remaining ones (see Exercise 8.2). Indeed, the same formula applies if the numbers concerned are arbitrary real numbers rather than cardinalities of sets:

Theorem 8.2 *Let real numbers a_J and b_J be given for each subset J of $N = \{1,\ldots,n\}$. Then the following are equivalent:*

(a) $a_J = \displaystyle\sum_{J \subseteq I \subseteq N} b_I$ for all $J \subseteq N$;

(b) $b_J = \displaystyle\sum_{J \subseteq I \subseteq N} (-1)^{|J \setminus I|} a_I$ for all $J \subseteq N$.

Proof The theorem asserts the form of the solution to a system of linear equations; in other words, the inverse of a certain matrix. However, the same matrix occurs in the original form of PIE.

The theorem as stated involves sums over supersets of the given index set. However, it is easily transformed to involve sums over subsets (see Exercise 8.3). In this form, it is a generalisation of the inverse relationship between the triangular matrix of binomial coefficients and the signed version (see Exercise 8.4).

8.2 Partially ordered sets

In this section, we formalise the kind of lower-triangular matrices which occurred in the last.

A *partial order* on a set X is a binary relation $\leq$ on X which satisfies the following conditions:

- $x \leq x$ *(reflexivity)*;
- if $x \leq y$ and $y \leq x$ then $x = y$ *(antisymmetry)*;
- if $x \leq y$ and $y \leq z$ then $x \leq z$ *(transitivity)*.

It is a *total order* if it satisfies the further condition

- for any x, y, exactly one of $x < y$, $x = y$, $y < x$ holds *(trichotomy)*, where $x < y$ is short for $x \le y$ and $x \ne y$. (Note that antisymmetry implies that at most one of these three conditions holds.)

The usual order relations on the natural numbers, integers, and real numbers are total orders. An important example of a partial order is the relation of *inclusion* on the set of all subsets of a given set. Other important examples of partially ordered sets include

- the positive integers ordered by divisibility (that is, $x \le y$ if and only if $x \mid y$);

- the subspaces of a finite vector space, ordered by inclusion. (This is known as a *projective space*.)

Any finite totally ordered set can be written as $\{x_1, x_2, \ldots, x_n\}$, where $x_i \le x_j$ if and only if $i \le j$.

A set carrying a partial order relation is called a *partially ordered set*, or *poset* for short.

We need to use the following result. A relation σ is an *extension* of a relation ρ if $x \rho y \Rightarrow x \sigma y$; that is, regarding a relation in the usual way as a set of ordered pairs, ρ is a subset of σ.

Theorem 8.3 *Any partial order on a set X can be extended to a total order on X.*

This theorem is easily proved for finite sets: take any pair of elements x, y which are incomparable in the given relation; set $x \le y$, and include all consequences of transitivity (show that no conflicts arise from this); and repeat until all pairs are comparable. It is more problematic for infinite sets; it cannot be proved from the Zermelo–Fraenkel axioms, but requires an additional principle such as the Axiom of Choice.

The upshot of the theorem for finite sets is that any finite partially ordered set can be written as $X = \{x_1, \ldots, x_n\}$ so that, if $x_i \le x_j$, then $i \le j$ (but not necessarily conversely). This is often possible in many ways. For example, the subsets of $\{a, b, c\}$, ordered by inclusion, can be written as

$$X_1 = \emptyset, \qquad X_2 = \{a\}, \qquad X_3 = \{b\}, \qquad X_4 = \{c\},$$
$$X_5 = \{a, b\}, \quad X_6 = \{a, c\}, \quad X_7 = \{b, c\}, \quad X_8 = \{a, b, c\}.$$

Now any function f from $X \times X$ to the real numbers can be written as an $n \times n$ matrix A_f, whose (i, j) entry is $f(x_i, x_j)$.

Our results extend to some infinite partially ordered sets, namely, those which are *locally finite*. (A partially ordered set X is locally finite if, for any $x, y \in X$, the *interval*

$$[x, y] = \{z \in X : x \le z \le y\}$$

is finite.)

Examples of infinite, locally finite posets include:

- The natural numbers; the integers (with the usual order).
- All finite subsets of an infinite set (ordered by inclusion).
- All finite-dimensional subspaces of an infinite-dimensional vector space over a finite field (ordered by inclusion).
- The positive integers (ordered by divisibility).

8.3   The incidence algebra of a poset

Let X be a partially ordered set. We can suppose that there is a total order on X extending the partial order, so that, if X is finite, then $X = \{x_1, \ldots, x_n\}$, so that if $x_i \leq x_j$, then $i \leq j$. Now we can represent a function $\alpha : X \times X \to \mathbb{R}$ as a matrix A_α with rows and columns indexed by X (or by $\{1, \ldots, n\}$), so that the (i, j) entry of A_α is $\alpha(x_i, x_j)$.

The *incidence algebra* of the partially ordered set X is defined to be the set of all functions $\alpha : X \times X \to \mathbb{R}$ which have the property that $\alpha(x, y) = 0$ unless $x \leq y$. Note that, for such a function α, the matrix A_α is lower triangular. The algebra operations of addition and multiplication are defined to be the usual matrix operations on the corresponding matrices; that is,

$$
\begin{aligned}
(\alpha + \beta)(x, y) &= \alpha(x, y) + \beta(x, y), \\
(\alpha\beta)(x, y) &= \sum_{x \leq z \leq y} \alpha(x, z)\beta(z, y).
\end{aligned}
$$

(These equations show that the way in which we extend the partial order to a total order does not affect the definitions.)

The definitions of addition and multiplication work equally well for an infinite locally finite poset (since the sum in the formula for multiplication is finite). So the incidence algebra of a locally finite poset is defined.

The incidence algebra has an identity, the function ι given by

$$
\iota(x, y) = \begin{cases} 1 & \text{if } x = y, \\ 0 & \text{otherwise.} \end{cases}
$$

(The matrix A_ι is the usual identity matrix.) Another important algebra element is the *zeta function* ζ, defined by

$$
\zeta(x, y) = \begin{cases} 1 & \text{if } x \leq y, \\ 0 & \text{otherwise.} \end{cases}
$$

Thus ζ is the characteristic function of the partial order, and an arbitrary function α belongs to the incidence algebra if and only if

$$
\zeta(x, y) = 0 \Rightarrow \alpha(x, y) = 0.
$$

A lower triangular matrix with ones on the diagonal has an inverse. The *Möbius function* μ of a poset is the inverse of the zeta function. In other words, it satisfies

$$\sum_{x \leq z \leq y} \mu(x,y) = \begin{cases} 1 & \text{if } x = y, \\ 0 & \text{otherwise.} \end{cases}$$

In particular, $\mu(x,x) = 1$ for all x. Moreover, if we know $\mu(x,z)$ for $x \leq z < y$, then we can calculate

$$\mu(x,y) = -\sum_{x \leq z < y} \mu(x,z).$$

In particular, we see that the values of the Möbius function are all integers.

8.4   Some Möbius functions

By definition, the Möbius function of a poset satisfies the following:

Proposition 8.4 *Let f and g be elements of the incidence algebra of a poset X (that is, functions on $X \times X$ satisfying $f(x,y) = g(x,y) = 0$ unless $x \leq y$). Then the following conditions are equivalent:*

(a) $g(x,y) = \displaystyle\sum_{x \leq z \leq y} f(x,z)$;

(b) $f(x,y) = \displaystyle\sum_{x \leq z \leq y} g(x,z)\mu(z,y)$.

This result is referred to as *Möbius inversion*. In order to use it, we have to compute the Möbius functions of various posets. Note that the Möbius function is local, in the sense that the value of $\mu(x,y)$ is determined by the structure of the interval $[x,y] = \{z : x \leq z \leq y\}$.

One important result is the following. Let $X_1, \ldots, X_r$ be posets. The *direct product* $X_1 \times \cdots \times X_r$ is the poset whose elements are all r-tuples $(x_1, \ldots, x_r)$ with $x_i \in X_i$ for $1 \leq i \leq r$; the order is given by

$$(x_1, \ldots, x_r) \leq (y_1, \ldots, y_r) \Leftrightarrow x_i \leq y_i \text{ for } 1 \leq i \leq r,$$

where the order $x_i \leq y_i$ is that in the poset X_i.

Proposition 8.5 *The Möbius function of the direct product $X_1 \times \cdots \times X_r$ is given by*

$$\mu((x_1, \ldots, x_r), (y_1, \ldots, y_r)) = \prod_{i=1}^{r} \mu_i(x_i, y_i),$$

where μ_i is the Möbius function of X_i.

Proof It is enough to show that

$$\sum_{\substack{x_i \leq z_i \leq y_i \\ 1 \leq i \leq r}} \prod_{i=1}^{r} \mu_i(x_i, z_i) = 0.$$

Now the left-hand side of this expression factorises as

$$\prod_{i=1}^{r} \sum_{x_i \leq z_i \leq y_i} \mu_i(x_i, z_i),$$

and the inner sum is zero by definition of the Möbius function μ_i.

Example: The integers In the poset of integers, with the usual order, the Möbius function is given by

$$\mu(x, y) = \begin{cases} 1 & \text{if } y = x; \\ -1 & \text{if } y = x + 1; \\ 0 & \text{otherwise.} \end{cases}$$

Example: Finite subsets of a set In this case, the Möbius function is

$$\mu(X, Y) = (-1)^{|Y \setminus X|} = (-1)^{|Y| - |X|} \text{ for } X \subseteq Y,$$

and of course $\mu(X, Y) = 0$ otherwise. For let $X \subseteq Y$, and let $Y \setminus X = \{z_1, \ldots, z_n\}$. We claim that the interval $[X, Y]$ is isomorphic to $\{0, 1\}^n$, the direct product of n copies of $\{0, 1\} \subseteq \mathbb{Z}$. The isomorphism takes a set Z with $X \leq Z \leq Y$ to the n-tuple $(e_1, \ldots, e_n)$, where

$$e_i = \begin{cases} 1 & \text{if } z_i \in Z, \\ 0 & \text{otherwise.} \end{cases}$$

So $\mu(X, Y)$ is equal to $\mu((0, \ldots, 0), (1, \ldots, 1))$ calculated in $\{0, 1\}^n$; by Proposition 8.5 this is $\mu(0, 1)^n$, and $\mu(0, 1) = -1$ by the preceding example.

Example: Positive integers ordered by divisibility Suppose that m divides n. Let $n/m = p_1^{a_1} p_2^{a_2} \cdots p_r^{a_r}$, where $p_1, \ldots, p_r$ are distinct primes and $a_1, \ldots, a_r$ positive integers. Then the interval $[m, n]$ is isomorphic to the direct product

$$[0, a_1] \times \cdots \times [0, a_r]$$

of intervals $[0, a_i]$ in $\mathbb{Z}$. The correspondence is given by

$$(b_1, \ldots, b_r) \leftrightarrow m p_1^{b_1} \cdots p_r^{b_r}.$$

By the first example, we see that $\mu(m, n) = 0$ if any $a_i > 1$, that is, if n/m is divisible by the square of a prime. If n/m is the product of s distinct primes, then $\mu(m, n) = (-1)^s$. To summarise:

$$\mu(m, n) = \begin{cases} (-1)^s & \text{if } n/m \text{ is the product of } s \text{ distinct primes;} \\ 0 & \text{if } m \text{ doesn't divide } n \text{ or if } n/m \text{ is not squarefree.} \end{cases}$$

Example: Subspaces of a finite vector space By the Second Isomorphism Theorem, if U and W are subspaces of V with $U \subseteq W$, then the interval $[U,W]$ is isomorphic to the poset of subspaces of W/U, and in particular depends only on $\dim(W) - \dim(U)$. It suffices to calculate $\mu(\{0\}, V)$, where V is an n-dimensional vector space over $\mathrm{GF}(q)$.

Now putting $x = -1$ in the q-binomial theorem, we obtain

$$\sum_{k=0}^{n} (-1)^k q^{k(k-1)/2} \begin{bmatrix} n \\ k \end{bmatrix}_q = 0$$

for $n > 0$. This is exactly the inductive step in the proof that

$$\mu(\{0\}, V) = (-1)^n q^{n(n-1)/2}$$

for $n > 0$. For there are $\begin{bmatrix} n \\ k \end{bmatrix}_q$ k-dimensional subspaces of V, and the induction hypothesis asserts that $\mu(\{0\}, W) = (-1)^k q^{k(k-1)/2}$ for each such subspace; then the identity shows that $\mu(\{0\}, V)$ must have the claimed value.

So, in general, $\mu(U, W) = (-1)^n q^{n(n-1)/2}$ if $U \subseteq W$ and $\dim(W/U) = n$; and of course, $\mu(U, W) = 0$ if $U \not\subseteq W$.

8.5 Classical Möbius inversion

All our examples in the preceding section have the special property that each interval $[x, y]$ is isomorphic to $[e, z]$, where e is a fixed element of the poset, and z depends on x and y. Thus, for the integers, $e = 0$ and $z = y - x$; for subsets of a set, $e = \emptyset$ and $z = y \setminus x$; for positive integers ordered by divisibility, $e = 1$ and $z = y/x$; and for subspaces of a vector space, $e = \{0\}$ and $z = y/x$ (the quotient space).

Thus, in these cases, the Möbius function satisfies $\mu(x, y) = \mu(e, z)$, so it can be written as a function of one variable z. Abusing notation, we use the same symbol μ. In the four cases, we have:

- $\mu(0) = 1$, $\mu(1) = -1$, $\mu(z) = 0$ for $z \geq 2$;

- $\mu(Z) = (-1)^{|Z|}$;

- $\mu(z) = (-1)^s$ if z is the product of s distinct primes, $\mu(z) = 0$ if z is not squarefree;

- $\mu(Z) = (-1)^k q^{k(k-1)/2}$, where $k = \dim(Z)$.

The third of these is the 'classical' Möbius function, and plays an important role in number theory. If you see $\mu(z)$ without any further explanation, it probably means the classical Möbius function. In this case, Möbius inversion can be stated as follows:

Proposition 8.6 *Let f and g be functions on the positive integers. Then the following are equivalent:*

(a) $g(n) = \sum_{m|n} f(m)$;

(b) $f(n) = \sum_{m|n} g(m)\mu(n/m)$.

Here are two applications of this result.

Example: Euler's function   Euler's ϕ-function (sometimes called the *totient function*) is the function ϕ defined on the positive integers by the rule that $\phi(n)$ is the number of integers x with $1 \le x < n$ coprime to n.

If $\gcd(x,n) = d$, then $\gcd(x/d, n/d) = 1$. So the number of x in this range with $\gcd(x,n) = d$ is $\phi(n/d)$, and we have

$$\sum_{d|n} \phi(n/d) = n,$$

or, putting $m = n/d$,

$$\sum_{m|n} \phi(m) = n.$$

Now Möbius inversion gives

$$\phi(n) = \sum_{m|n} m\mu(n/m).$$

From this it is easy to deduce that, if $n = p_1^{a_1} \cdots p_r^{a_r}$, where p_i are distinct primes and $a_i > 0$, then

$$\phi(n) = p_1^{a_1-1}(p_1 - 1) \cdots p_r^{a_r-1}(p_r - 1).$$

Example: Irreducible polynomials   Let $f_q(n)$ be the number of monic irreducible polynomials of degree n over GF(q). By Theorem 6.9,

$$\sum_{m|n} m f_q(m) = q^n.$$

So, by Möbius inversion, we have a formula for $f_q(n)$:

$$f_q(n) = \frac{1}{n} \sum_{m|n} q^m \mu(n/m).$$

For example, the number of irreducible polynomials of degree 6 over GF(2) is

$$\frac{1}{6}(2^6 - 2^3 - 2^2 + 2^1) = 9.$$

(Why is the word 'monic' not needed here?)

8.6 The general linear group

The set of all permutations of $\{1,\ldots,n\}$ forms a group, the *symmetric group*, with order $n!$. We saw in Chapter 6 that the q-analogue of this group is the *general linear group* $GL(n,q)$, the group of all bijective linear transformations of a vector space V of dimension n over $GF(q)$ (where q is a prime power).

We saw in Proposition 6.8 that the order of $GL(n,q)$ is given by the two formulae

$$(q^n - 1)(q^n - q)\cdots(q^n - q^{n-1}) = (-1)^n q^{n(n-1)/2} \sum_{k=0}^n (-1)^k q^{k(k+1)/2} \begin{bmatrix} n \\ k \end{bmatrix}_q.$$

Here is a different proof based on Möbius inversion.

Let V be an n-dimensional vector space over $GF(q)$. For any subspace W of V of dimension $n - k$, the number of linear maps on V whose range is contained in W is equal to $q^{n(n-k)}$. For such a map is uniquely determined by the images of the n vectors of a basis for q; each of these images can be any of the q^{n-k} vectors of W; so there are altogether $(q^{n-k})^n$ choices of the linear map.

Let $a(W)$ be the number of linear maps whose image is precisely W. Clearly if $\dim(W) = n - k$, then we have

$$\sum_{X \subseteq W} a(X) = q^{n(n-k)},$$

since if the image of a map is contained in W then it is equal to some unique subspace of W. So the principle of Möbius inversion gives

$$\sum_{X \subseteq W} q^{n(n-l)} \mu(X, W) = a(W),$$

where $l = \dim(X)$. Taking $W = V$, and replacing the dummy variables X and l by W and k we obtain

$$\begin{aligned}
a(V) &= \sum_{W \subseteq V} q^{n(n-k)} (-1)^k q^{k(k-1)/2}, \\
&= \sum_{k=0}^n q^{n(n-k)} \begin{bmatrix} n \\ k \end{bmatrix}_q (-1)^k q^{k(k-1)/2}, \\
&= \sum_{k=0}^n q^{nk} \begin{bmatrix} n \\ k \end{bmatrix}_q (-1)^{n-k} q^{(n-k)(n-k-1)/2},
\end{aligned}$$

since $\mu(W, V) = (-1)^k q^{k(k-1)/2}$ if $\dim(V) - \dim(W) = k$. (We have also used the fact that $\begin{bmatrix} n \\ k \end{bmatrix}_q = \begin{bmatrix} n \\ n-k \end{bmatrix}_q$.) Now $a(V)$ is the required number of bijective linear maps; a little manipulation gives the required formula. The exponent of q in the kth term is

$$nk + \frac{(n-k)(n-k-1)}{2} = \frac{n(n-1)}{2} + \frac{k(k+1)}{2}.$$

8.7 Exercises

8.1 Before doing this exercise, you should review Exercise 3.12.

(a) In how many ways can k identical sweets be given to n children if we require that the number x_i of sweets given to the ith child should satisfy $x_i \leq b_i$, for some numbers $b_1, \ldots, b_n$?

(b) In how many ways can k identical sweets be given to n children if we require that the number x_i of sweets given to the ith child should satisfy $a_i \leq x_i \leq b_i$, for some numbers $a_1, b_1, \ldots, a_n, b_n$?

(In this exercise, you are not required to write down an explicit formula; description of a method for calculating the number will suffice.)

8.2 Let $A_1, \ldots, A_n$ be subsets of X. For $J \subseteq N = \{1, \ldots, n\}$, let A_j consist of the points of X lying in A_i for all $i \in J$, and B_j the points lying in A_i if $i \in J$ and not if $i \notin J$. Show that

$$|B_J| = \sum_{J \subseteq K \subseteq N} (-1)^{|K \setminus J|} |A_K|.$$

8.3 Prove that, with the hypotheses of Theorem 8.2, the following conditions are equivalent:

(a) $a_J = \sum_{I \subseteq J} b_I$ for all $J \subseteq N$;

(b) $b_J = \sum_{I \subseteq J} (-1)^{|J \setminus I|} a_I$ for all $J \subseteq N$.

8.4 By taking the numbers a_J and b_J of the preceding exercise to depend only on the cardinality j of J, show that the following statements are equivalent for two sequences (x_i) and (y_i):

(a) $x_j = \sum_{i=0}^{j} \binom{j}{i} y_i$;

(b) $y_j = \sum_{i=0}^{j} (-1)^{j-i} \binom{j}{i} y_i$.

8.5 Prove that

$$S(n,k) = \frac{1}{k!} \sum_{j=0}^{k} (-1)^j \binom{k}{j} (k-j)^n.$$

8.6 Prove that the solution M_n to the *problème des ménages* satisfies the recurrence

$$(n-2)M_n = n(n-2)M_{n-1} + nM_{n-2} + 4(-1)^{n-1}$$

for $n \geq 4$.

8.7 Let x and y be elements of a poset X, with $x \leq y$. A *chain* from x to y is a sequence $x = x_0, x_1, \ldots, x_l = y$ with $x_{i-1} < x_i$ for $i = 1, \ldots, l$; its *length* is l. Show that

$$\mu(x, y) = \sum_{c \in C} (-1)^{l(c)},$$

where C is the set of all chains from x to y, and $l(c)$ is the length of c.

8.8 Let $d(n)$ be the number of divisors of the positive integer n. Prove that

$$\sum_{m|n} d(m)\mu(n/m) = 1$$

for $n > 1$.

8.9 Let $\mathscr{P}(X)$ denote the poset whose elements are the partitions of the set X, with $P \leq Q$ if P refines Q (that is, every part of P is contained in a part of Q). Let E be the partition into sets of size 1.

(a) Show that, if the parts of P have sizes $a_1, \ldots, a_r$, then

$$\mu(E, P) = (a_1 - 1)! \cdots (a_r - 1)!.$$

(b) Show that any interval $[P, Q]$ is isomorphic to a product of posets of the form $\mathscr{P}(X_j)$, and hence calculate $\mu(P, Q)$.

8.10 Let G be the cyclic group consisting of all powers of the permutation

$$g : 1 \rightarrow 2 \rightarrow \cdots \rightarrow n \rightarrow 1.$$

Show that the cycle index of G is

$$Z(G) = \frac{1}{n} \sum_{m|n} \phi(n/m) s_{n/m}^m,$$

where ϕ is Euler's function.

8.11 A necklace is made of n beads of q different colours. Necklaces which differ only by a rotation are regarded as the same. Show that the number of different necklaces is

$$\frac{1}{n}\sum_{m|n} q^m \phi(n/m),$$

while the number which have no rotational symmetry is

$$\frac{1}{n}\sum_{m|n} q^m \mu(n/m).$$

(Notice that, if q is a prime power, the second expression is equal to the number of monic irreducible polynomials of degree n over $GF(q)$. Finding a bijective proof of this fact is much harder!)

8.12 A function F on the natural numbers is said to be *multiplicative* if

$$\gcd(m,n) = 1 \Rightarrow F(mn) = F(m)F(n).$$

(a) Suppose that F and G are multiplicative. Show that the function H defined by

$$H(n) = \sum_{k|n} F(k)G(n/k)$$

   is multiplicative.

(b) Show that the Möbius and Euler functions are multiplicative.

(c) Let $d(n)$ be the number of divisors of n, and $\sigma(n)$ the sum of the divisors of n. Show that d and σ are multiplicative.

8.13 The following problem, based on the children's game 'Screaming Toes', was suggested to me by Julian Gilbey.

> n people stand in a circle. Each player looks down at someone else's feet (i.e., not at their own feet). At a given signal, everyone looks up from the feet to the eyes of the person they were looking at. If two people make eye contact, they scream. What is the probability of at least one pair of people screaming?

Prove that the required probability is

$$\sum_{k=1}^{\lfloor n/2 \rfloor} \frac{(-1)^{k-1}(n)_{2k}}{2^k k!(n-1)^{2k}},$$

where $(n)_j = n(n-1)\cdots(n-j+1)$.
   How does this function behave as $n \to \infty$?

8.14 Prove the following 'approximate version' of PIE:

Let $A_1,\ldots,A_n,A'_1,\ldots,A'_n$ be subsets of a set X. For $I \subseteq N = \{1,\ldots,n\}$, let

$$a_I = \left|\bigcap_{i\in I} A_i\right|, \qquad a'_I = \left|\bigcap_{i\in I} A'_i\right|.$$

If $a_I = a'_I$ for all *proper* subsets I of N, then $|a_N - a'_N| \leq |X|/2^{n-1}$.

Remark For more general approximate versions of PIE, see the paper by Linial and Nisan in the bibliography.

8.15 Let $n = k_1 + k_2 + \cdots + k_r$, where $k_1,\ldots,k_r$ are positive integers, and suppose that the poset P on $\{1,\ldots,n\}$ consists of r disjoint chains of lengths $k_1,\ldots,k_r$. Show that the number of total orders extending P is the multinomial coefficient

$$\binom{n}{k_1,k_2,\ldots,k_r}.$$

CHAPTER 9

The Tutte polynomial

In this chapter, we examine a polynomial associated with a graph, or more generally with a matroid, which encodes a great deal of counting information. This polynomial, the Tutte polynomial, also has close connections with statistical mechanics: it is, up to a trivial transformation, the partition function of the Potts model on the graph in question. We also look at the problem of producing an 'orbit-counting' version of the Tutte polynomial.

9.1 The chromatic polynomial

In combinatorics, the first question we ask is whether a certain arrangement is possible; if it is, then we ask in how many ways it can be done. In the case of the chromatic polynomial, the procedure was reversed; Birkhoff hoped that by counting the number of arrangements and proving that it is non-zero, he could show that the required arrangement exists.

The famous *Four Colour Conjecture* asked whether every map drawn in the plane can be coloured with four colours in such a way that countries sharing a border are given different colours. It is straightforward to convert this into a problem about planar graphs (those which can be drawn in the plane without edges crossing) and vertex-colourings: simply represent each country by a vertex, and if two countries share a border, draw an edge joining their vertices crossing this border. Then the requirement is that vertices joined by an edge should have two different colours.

Let Γ be any graph, and let $P_\Gamma(q)$ denote the number of ways of colouring the vertices of Γ with q colours $c_1, \ldots, c_q$ in such a way that adjacent vertices have different colours. (Such a colouring is called *proper*.) If Γ has a loop (an edge joining a vertex to itself), then clearly no proper colouring exists.

Proposition 9.1 *For a fixed (finite) loopless graph Γ with n vertices, the function $q \mapsto P_\Gamma(q)$ is a monic polynomial of degree n.*

Proof We can prove this, and find a formula for the polynomial, using the Principle of Inclusion and Exclusion, which we discussed in Chapter 8. Let E be the set of edges, and X the set of all colourings (proper or not) of the vertices of the graph with a fixed set of q colours. If f is a colouring, call an edge *bad* if its two vertices have the same colour. In how many colourings does a given subset F of E consist of bad edges?

Let $\Gamma(F)$ be the graph with the same vertex set as Γ and with edge set F. In any colouring in which the edges of F are bad, then any two vertices in the same connected component of $\Gamma(F)$ have the same colour; so the number of such colourings is $q^{c(F)}$, where $c(F)$ denotes the number of connected components of $\Gamma(F)$.

So by PIE, the number of colourings with no bad edges is

$$P_\Gamma(q) = \sum_{F \subseteq E} (-1)^{|F|} q^{c(F)}. \tag{9.1}$$

Now the maximum number of connected components of $\Gamma(F)$ is n (the number of vertices of Γ), occurring only in the case where $F = \emptyset$. So the displayed formula is a polynomial with leading term q^n.

The polynomial P_Γ is called the *chromatic polynomial* of Γ.

The more common approach to the chromatic polynomial uses the operations of deletion and contraction of a graph. Let Γ be a graph, and $e = \{v, w\}$ an edge of Γ.

- The graph $\Gamma \backslash e$ obtained by *deletion* of e is simply the graph obtained by removing e from the edge set of Γ.

- The graph Γ / e obtained by *contraction* of e is the graph obtained by removing the edge e and then identifying the vertices v and w. (If there is a vertex x joined to both v and w, this results in a double edge from x to the new vertex.)

Lemma 9.2 *If e is an edge of the graph Γ, then*

$$P_\Gamma(q) = P_{\Gamma \backslash e}(q) - P_{\Gamma / e}(q).$$

Proof Let $e = \{v, w\}$. Consider the set of colourings of $\Gamma \backslash e$, and partition this set into two subsets:

- colourings f with $f(v) = f(w)$: each of these induces, in a natural way, a colouring of Γ/e.
- colourings f with $f(v) \neq f(w)$: these are proper colourings of Γ.

So $P_{\Gamma \backslash e}(q) = P_{\Gamma/e}(q) + P_{\Gamma}(q)$, as required.

By successive deletion and contraction (together with the removal of multiple edges if they occur – this does not affect the number of proper colourings) we are reduced to a collection of graphs with no edges. The number of proper colourings of a graph with n vertices and no edges is q^n.

This gives a method of computing the chromatic polynomial (not very efficient, since exponentially many steps are required), and also shows that, if the chromatic polynomial is computed by deletion and contraction, the result is independent of the order in which the edges are chosen in the algorithm. It gives another proof that the result is a polynomial in q.

Birkhoff hoped that it would be possible to show that the chromatic polynomial $P_{\Gamma}(x)$ of a planar graph Γ cannot have a root at $x = 4$. This would show that the number of colourings of such a graph with four colours could not be zero, and prove the Four Colour Conjecture. However, nobody has succeeded in doing this, and the proof of the conjecture by Appel and Haken used completely different techniques.

Richard Stanley gave a counting interpretation of the values of the chromatic polynomial at negative integers. Here is an example. An *acyclic orientation* of a graph Γ is an assignment of directions to the edges of Γ in such a way that no directed cycles are created. For example, there are 2^n orientations of an n-cycle, of which just two fail to be acyclic.

Theorem 9.3 *The number of acyclic orientations of a graph Γ on n vertices is* $(-1)^n P_{\Gamma}(-1)$.

Proof We use the deletion-contraction formula. Choose an edge e, and consider the set S of acyclic orientations of $\Gamma \backslash e$.

Observe that every acyclic orientation of $\Gamma \backslash e$ can be extended to an acyclic orientation of Γ. This could only fail if both choices of orientation of e lead to directed cycles. But, as the figure shows, this would mean that there was a directed cycle in $\Gamma \backslash e$:

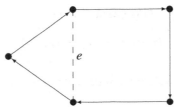

Let S be the set of acyclic orientations of $\Gamma \backslash e$. We divide S into two classes:

- S_1 is the set of orientations which force the direction of e, because there is a directed path in some direction between its endpoints.

- S_2 consists of orientations in which either ordering of S will be admissible (will create no directed cycle).

Thus the number of acyclic orientations of Γ is $|S_1| + 2|S_2|$.

Moreover, any orientation in the class S_2 gives rise to an acyclic orientation of Γ/e, since by assumption, identifying the ends of e does not create a directed cycle. Also $|S_1| + |S_2|$ is the number of acyclic orientations of $\Gamma \backslash e$.

By induction, we may assume that $|S_1| + |S_2| = (-1)^n P_{\Gamma \backslash e}(-1)$ and $|S_2| = (-1)^{n-1} P_{\Gamma/e}(-1)$. So the number of acyclic orientations of Γ is given by

$$
\begin{aligned}
|S_1| + 2|S_2| &= (|S_1| + |S_2|) + |S_2| \\
&= (-1)^n P_{\Gamma \backslash e}(-1) + (-1)^{n-1} P_{\Gamma/e}(-1) \\
&= (-1)^n P_{\Gamma}(-1),
\end{aligned}
$$

and the proof is complete. (In the penultimate line we use the deletion-contraction formula for the chromatic polynomial of Γ, namely $P_\Gamma = P_{\Gamma \backslash e} - P_{\Gamma/e}$.)

The induction is on the number of edges, so we start the induction with null graphs. The null graph on n vertices has chromatic polynomial q^n and has a single acyclic orientation, and $(-1)^n(-1)^n = 1$ as required.

9.2 The Tutte polynomial

Tutte generalised the notion of the chromatic polynomial to a two-variable polynomial which is now called the Tutte polynomial. Sokal considered a more general polynomial which has one local variable associated with each edge of the graph, together with one global variable. In statistical mechanics, the local variables represent some measures of the strength of interaction associated with each edge, while the global variable is something like the temperature of the system. The usual Tutte polynomial is obtained when all the local variables are equal. Sokal's paper contains much more detailed and accurate information about this.

Sokal has given convincing reasons for using the multivariate polynomial where possible. However, for reasons that will appear in the next section, I will stick to the usual two-variable polynomial. Also, both the two-variable and the multivariable forms extend naturally to matroids, which provide the correct level of generality; but it would take us too far afield to consider this, so I will deal only with graphs.

The following definition is due to Whitney; Tutte's definition was more complicated, though equivalent. Let Γ be a graph with n vertices having edge set E.

For any subset F of E, define the *rank* of F to be $r(F) = n - c(F)$, where as before $c(F)$ is the number of connected components of the graph with edge set F. Said otherwise, $r(F)$ is the size of the largest subgraph of $\Gamma(F)$ which is a forest (that is, contains no cycles). Then the *Tutte polynomial* of Γ is

$$T(\Gamma;x,y) = \sum_{F \subseteq E} (x-1)^{r(E)-r(F)}(y-1)^{|F|-r(F)}.$$

The next theorem gives some specialisations of the Tutte polynomial. A *spanning subgraph* of Γ is a graph having all the vertices and some of the edges of Γ; that is, of the form $\Gamma(F)$ for some $F \subseteq E$. It is called a *spanning tree* or *spanning forest* if it is a tree or forest respectively.

Proposition 9.4 *Let Γ be a connected graph.*

(a) $T(\Gamma;1,1)$ *is the number of spanning trees of Γ.*

(b) $T(\Gamma;2,1)$ *is the number of spanning forests of Γ.*

(c) $T(\Gamma;1,2)$ *is the number of connected spanning subgraphs of Γ.*

(d) $T(\Gamma;2,2) = 2^{|E|}$.

(e) $qT(\Gamma;1-q,0)$ *is the chromatic polynomial of Γ.*

Proof (a) Putting $x = y = 1$ selects only those terms in which the exponent of both $x - 1$ and $y - 1$ is zero, that is, corresponding to subsets F for which $|F| = r(F) = r(E)$. The second inequality asserts that $\Gamma(F)$ is connected and the first that it is a forest; so only trees are selected. Each tree contributes 1 to the evaluation.

(b) and (c) Similarly, putting $x = 2$ and $y = 1$ selects forests, while putting $x = 1$ and $y = 2$ selects connected graphs.

(d) is trivial since every term in the sum is 1.

(e) If $x = 1 - q$ and $y = 0$, then the term corresponding to F in the sum is

$$(-q)^{r(E)-r(F)}(-1)^{|F|-r(F)} = q^{n-1-r(F)}(-1)^{|F|} = q^{c(F)-1}(-1)^{|F|},$$

using the fact that $r(F) = n - c(F)$. Comparing with the Inclusion-Exclusion formula (9.1) gives the result.

We need to define a couple more notions. Let A be an abelian group, written additively. Choose an arbitrary orientation of the edges of Γ. A function f from oriented edges to A is called a *flow* if, for any vertex v, the sum of the values of f on edges entering v is equal to the sum of the values on edges leaving v (calculated in the group A). In other words, the algebraic sum of f over all the edges incident with v is zero, where the values on edges leaving v are given a $-$ sign. Dually, a function f from oriented edges to A is a *tension* if the algebraic sum of the values of f around any circuit in Γ is zero (where edges in the opposite direction to the circuit are given a $-$ sign). Note that, if we reverse the orientation of an edge, and change the sign of the function f on that edge, the properties of being a flow or a tension are unaltered. So the numbers of flows and tensions are independent of the chosen orientation of the edges.

A flow or tension is *nowhere-zero* if it does not take the zero value (the identity of the group A).

Proposition 9.5 *Let A be an abelian group of order q. Then the number of nowhere-zero tensions in a graph Γ with values in A is $T(\Gamma; 1-q, 0)$, while the number of nowhere-zero flows in Γ with values in A is $T(\Gamma; 0, 1-q)$.*

Remark Note that these numbers do not depend on the structure of the group A, but only on its order.

Proof Both facts can be proved by an Inclusion-Exclusion argument similar to that used in Proposition 9.1. But here is an alternative proof of the result about tensions. This argument 'explains' why the number of nowhere-zero tensions does not depend on the structure of A.

Orient the edges of Γ. First, choose any colouring f of Γ with q colours. Identify the set of colours with the abelian group A. Now let ∂f be the function from oriented edges of Γ to A given by $\partial f(e) = f(v) - f(w)$ if e is directed from w to v. It is easy to see that this is a tension; it is nowhere-zero if and only if the colouring is proper.

Conversely, given a nowhere-zero tension g, and two vertices v_0 and v, the algebraic sum of the values of g along any path from v_0 to v is the same. Now choose one vertex in each connected component of Γ, and define the function f arbitrarily at these vertices. For any vertex v, let v_0 be the chosen vertex in its component, and let $f(v)$ be the sum of $f(v_0)$ and the algebraic sum of the values of g on any path from v_0 to v. Then f is a proper colouring and $\partial f = g$.

So the number of proper colourings is $q^{c(E)}$ times the number of nowhere-zero tensions.

There are many other counts associated with a graph, whose values are evaluations of the Tutte polynomial. For example, Stanley's Theorem 9.3 shows that the number of acyclic orientations of a graph Γ is equal to $T(\Gamma; 0, 2)$.

9.3 Orbit counting and the Tutte polynomial

Let us return briefly to the table in Chapter 3 of the number of ways of choosing k objects from a set of n, under different sampling protocols. Here I have rewritten the table to describe choosing n objects from a set of q, for reasons which will appear below.

	Order significant	Order not significant
With replacement	q^n	$\dbinom{q+n-1}{n}$
Without replacement	$(q)_n$	$\dbinom{q}{n}$

We can regard such a sample in which order is significant as a colouring of a graph on n vertices with a set of q colours: for the first, second, ..., nth vertex, we have to choose a colour. Sampling with replacement gives an arbitrary colouring, which is of course a proper colouring of the null graph (the graph with no edges). Sampling without replacement gives a colouring in which all colours are distinct; this is a proper colouring of the complete graph (in which all pairs of vertices are joined by an edge).

A sample in which order is not significant gives rise to a colouring of the null or complete graph, where we count two colourings as the same if some permutation of the vertices carries the first to the second. In other words, for sampling with order not significant, we are counting orbits of the symmetric group on proper colourings of the null or complete graph.

Noting that the automorphism group of either of these graphs is the symmetric group, we can now ask:

> Given a graph Γ and a group G consisting of automorphisms of Γ, what is the number of orbits of G on proper colourings of Γ with q colours?

This question was considered in the paper by Cameron, Jackson and Rudd in the bibliography, in which further details are given. Here we first consider the orbital chromatic polynomial, which solves the question as stated, and then turn briefly to the (more general) orbital Tutte polynomial.

Let Γ be a graph and G a group of automorphisms of Γ (a subgroup of $\mathrm{Aut}(\Gamma)$). Let $P_{\Gamma,G}(q)$ be the number of G-orbits on proper colourings of Γ with q colours. By the Orbit-counting Lemma, we have

$$P_{\Gamma,G}(q) = \frac{1}{|G|} \sum_{g \in G} \phi(\Gamma, g, q),$$

where $\phi(\Gamma, g, q)$ is the number of proper colourings of Γ with q colours fixed by the automorphism g. Now a colouring of Γ is fixed by g if and only if it is constant on the cycles of g (in its action on vertices). So, if any cycle of g contains an edge of Γ, there are no such colourings; otherwise the number is $P_{\Gamma/g}(q)$, where Γ/g denotes the graph obtained by shrinking every cycle of g to a single vertex. If we make the convention that, if a cycle of g contains an edge, then Γ/g has a loop on the vertex resulting from shrinking this cycle, we conclude the following:

Proposition 9.6 *Let G be a group of automorphisms of the graph Γ. Then*

$$P_{\Gamma,G}(q) = \frac{1}{|G|} \sum_{g \in G} P_{\Gamma/g}(q).$$

In particular, if Γ is loopless, then the function $q \mapsto P_{\Gamma,G}(q)$ is a polynomial of degree n (the number of vertices of Γ) with leading coefficient $1/|G|$.

The orbital chromatic polynomial is a specialisation of an object called the *orbital Tutte polynomial*. Its definition requires some algebraic background, which I now outline.

An *integral domain* is a commutative ring with identity, in which there are no zero-divisors; that is, if $ab = 0$, then $a = 0$ or $b = 0$. A *principal ideal domain* is an integral domain in which every ideal is generated by a single element; that is, every set of elements has a greatest common divisor which can be expressed as a linear combination of elements of the set. Readers unfamiliar with these notions will lose nothing by replacing the words 'a principal ideal domain' by 'the ring $\mathbb{Z}$ of integers' in everything that follows. However, the generality is useful for other applications such as coding theory, where the ring is a Galois field.

As usual, we say that a divides b if $b = ac$ for some element c, and write this as $a \mid b$.

In an integral domain, a *unit* is an element which has a multiplicative inverse; two elements a and b are *associates* if $b = ua$ for some unit u. Being associates is an equivalence relation on the ring: two elements are associates if and only if each divides the other. In the ring of integers, the only units are 1 and -1; so the associate classes have the form $\{n, -n\}$ for non-negative integers n.

The next theorem talks about row and column operations on a matrix over a ring. These are similar to the familiar operations over a field, but with one change. We are permitted to add any multiple of a row (or column) to another; to multiply a row (or column) by *a unit*; and to interchange two rows (or columns). The reason for the restriction to units is that each operation should be undone by another operation of the same kind.

Theorem 9.7 *Let A be a matrix over a principal ideal domain R, and suppose that A has rank r. Then A can be converted, by row and column operations, into a matrix D in which the first r elements $d_1, \ldots, d_r$ on the main diagonal are non-zero and satisfy $d_i \mid d_{i+1}$ for $i = 1, \ldots, r-1$; all other entries in the matrix are zero. The elements $d_1, \ldots, d_r$ are unique up to being replaced by associates.*

The *Smith normal form* of A is the matrix D of the above theorem. The elements $d_1, \ldots, d_r$ and 0 ($n - r$ times) are the *invariant factors* of A.

We can represent an $m \times n$ matrix A of rank r by an *indicator monomial* $x(A)$, as follows. Let F be an index set for the set of associate classes in R. (If $R = \mathbb{Z}$, we can choose I to be the set of non-negative integers, where the integer i represents the class $\{i, -i\}$.) We take a family $(x_i : i \in I)$ of indeterminates indexed by I. Now if the non-zero elements of the Smith normal form of A belong to associate classes with indices $i_1, \ldots, i_r$, then $x(A)$ is the monomial $x_{i_1} \cdots x_{i_r} x_0^{n-r}$, where 0 indexes the associate class $\{0\}$. We need also the indicator monomial $y(A)$ in variables $(y_i : i \in I)$, defined in the same way.

Now we begin constructing the orbital Tutte polynomial of a graph Γ. Choose arbitrarily a direction of each edge of Γ, and a cyclic orientation of each circuit in Γ (independent of the directions of its edges). These orientations are required to define the polynomial, but the result is independent of the choices. The signed vertex-edge incidence matrix of Γ is the matrix M with rows indexed by vertices and columns by edges, with (v, e) entry

$$M_{ve} = \begin{cases} +1 & \text{if directed edge } e \text{ enters vertex } v, \\ -1 & \text{if directed edge } e \text{ leaves vertex } v, \\ 0 & \text{if } v \text{ is not incident with } e. \end{cases}$$

Similarly, we define the signed circuit-edge incidence matrix M^*, with rows indexed by circuits and columns by edges, with (C, e) entry

$$M_{Ce}^* = \begin{cases} +1 & \text{if } e \in C \text{ and the orientations agree}, \\ -1 & \text{if } e \in C \text{ and the orientations disagree}, \\ 0 & \text{if } e \notin C. \end{cases}$$

Finally, if g is an automorphism of Γ, then we define a signed edge-permutation matrix P_g with rows and columns indexed by edges, with

$$(P_g)_{ef} = \begin{cases} +1 & \text{if } e^g = f \text{ and the orientations of } e^g \text{ and } f \text{ agree}, \\ -1 & \text{if } e^g = f \text{ and the orientations of } e^g \text{ and } f \text{ disagree}, \\ 0 & \text{if } e^g \neq f. \end{cases}$$

We let M_g and M_g^* be the matrices (in block form)

$$M_g = \begin{pmatrix} M \\ P_g - I \end{pmatrix}, \qquad M_g^* = \begin{pmatrix} M^* \\ P_g - I \end{pmatrix}.$$

We need one more piece of notation. If M is a matrix with columns indexed by E, and $F \subseteq E$, then $M(F)$ denotes the matrix obtained by selecting just the columns with index in F.

Now the orbital Tutte polynomial is defined as follows. Let Γ be a graph with edge set E, and G a group of automorphisms of Γ. Let (x_i) and (y_i) be two families of variables indexed by the non-negative integers. Set

$$OT(\Gamma, G; x_i, y_i) = \frac{1}{|G|} \sum_{g \in G} \sum_{F \subseteq E} x(M_g(F)) y(M^*(E \setminus F)),$$

a polynomial in the variables x_i and y_i for $i \geq 0$.

Of course, since it is a polynomial, only finitely many of the variables occur. The following can be shown:

Proposition 9.8 *Let G be a group of automorphisms of the graph Γ. If either x_i or y_i is present in the polynomial $OT(\Gamma, G)$, then either $i = 0$ or there is an element of order i in the group G.*

We now show that suitable specialisations of the orbital Tutte polynomial count orbits of G on nowhere-zero flows and tensions, as well as on proper colourings. If A is a finite abelian group and m a non-negative integer, we let $\alpha_m(A)$ denote the number of solutions of $mx = 0$ in A. Note that $\alpha_0(A) = |A|$ and $\alpha_1(A) = 1$.

Theorem 9.9 *Let Γ be a graph, and G a group of automorphisms of Γ; let A be an abelian group of order q.*

(a) *If Γ is connected, then the number $P_{\Gamma, G}(q)$ of G-orbits on proper colourings of Γ with q colours is obtained from $OT(\Gamma, G)$ by substituting $x_i \leftarrow -1$, $y_0 \leftarrow q$, $y_i \leftarrow 1$ for $i > 0$, and multiplying by q.*

(b) *The number of G-orbits on nowhere-zero tensions with values in A is obtained from $OT(\Gamma, G)$ by substituting $x_i \leftarrow -1$, $y_i \leftarrow \alpha_i(A)$.*

(c) *The number of G-orbits on nowhere-zero flows with values in A is obtained from $OT(\Gamma, G)$ by substituting $x_i \leftarrow \alpha_i(A)$, $y_i \leftarrow -1$.*

Remark The substitutions in (a) and (b) are not the same; that in (b) depends on the structure of A, so the numbers of orbits on nowhere-zero flows or tensions do depend on the structure of A if the group G is not the identity.

Proof We give a sketch of the proof for flows. (The version for colourings follows from our earlier formula for the orbital chromatic polynomial, while the version for tensions is very similar to that for flows.)

First, let f be a column vector of elements of A, of length equal to the number m of columns of the integer matrix C. How many solutions of $Cf = 0$ are there? Row and column operations don't change the number of solutions, so we can assume that C is in Smith normal form, with non-zero elements $d_1, \ldots, d_r$. Then the equations become $d_1 f_1 = 0, \ldots, d_r f_r = 0$, with $d_{r+1}, \ldots, d_m$ arbitrary. So the number of solutions is $\alpha_{d_1}(A) \cdots \alpha_{d_r}(A) \alpha_0(A)^{m-r}$, which is just the indicator monomial $x(C)$ with $\alpha_i(A)$ substituted for x_i for all i (which we write $x_i \leftarrow \alpha_i(A)$).

The equations $Mf = 0$ assert that f is a flow, while the equations $(P_g - I)f = 0$ assert that f is fixed by g. So the number of flows fixed by g is $x(M_g)$, with $x_i \leftarrow \alpha_i(A)$ for all i.

Inclusion-Exclusion shows that the number of fixed nowhere-zero flows is

$$\sum_{F \subseteq E} (-1)^{|E \setminus F|} x(M_g(F)) : x_i \leftarrow \alpha_i(A)).$$

This is the polynomial

$$\sum_{F \subseteq E} x(M_g(F)) y(M_g^*(E \setminus F))$$

with the substitution $x_i \leftarrow \alpha_i(A)$, $y_i \leftarrow -1$ for all i.

Averaging over $g \in G$ and using the Orbit-counting Lemma gives the result.

It is known that the matrix M is *totally unimodular*; this means that any sub-determinant is equal to 0 or ± 1. It follows that all the invariant factors are 0 or 1, so that $x(M) = x_0^{m-r} x_1^r$ (where r is the rank of M). Since the row space of M^* is the null space of M, and *vice versa*, a similar assertion holds for M^*. So, if G is the trivial group, the orbital Tutte polynomial $OT(\Gamma, G)$ involves the variables x_0, x_1, y_0, y_1 only. This is also an immediate consequence of Proposition 9.8.

It is now not difficult to show that the substitution $x_1 \leftarrow 1$, $y_1 \leftarrow 1$, $x_0 \leftarrow y - 1$, $y_0 \leftarrow x - 1$ produces the usual Tutte polynomial of Γ.

Since the substitutions in parts (b) and (c) of Theorem 9.9 replace each of the variables x_0, x_1, y_0, y_1 by ± 1 or $|A|$, we see again that the number of nowhere-zero flows or tensions on Γ with values in A depends only on the order of A and not its structure.

9.4   The Matrix-Tree Theorem

It is known that evaluation of the Tutte polynomial in general is hard. But there are some specific points (x, y) in the plane for which it is easy to calculate $T(\Gamma; x, y)$. For a trivial example, $T(\Gamma; 2, 2) = 2^m$, where m is the number of edges of Γ.

A less trivial example is $T(\Gamma; 1, 1)$ which, as we have seen, is the number of spanning trees of Γ (if Γ is connected). This can be computed by linear algebra. The following result, known as the Matrix-Tree Theorem, was essentially proved by Kirchhoff in 1847. Kirchhoff, a physicist, was studying properties of electrical networks.

First we need a definition. Let Γ be a graph with vertex set $\{v_1, \ldots, v_n\}$. (Multiple edges but not loops are permitted.) The *Laplacian matrix* of Γ is the $n \times n$ matrix $L(\Gamma)$ whose (i, j) entry is equal to the number of edges incident with vertex i if $i = j$, and the negative of the number of edges joining i to j if $i \neq j$. Note that $L(\Gamma)$ is symmetric, and every row or column sum is zero (since, in the ith row, every edge incident with v_i contributes $+1$ to the diagonal entry and -1 to the (i, j) entry if its other end is v_j. So $L(\Gamma)$ is singular, and has determinant 0.

Theorem 9.10 (Matrix-Tree Theorem) *Let Γ be a connected graph. Then the following three quantities are equal:*

  (a) the number of spanning trees of Γ;

  (b) $(\lambda_2 \cdots \lambda_n)/n$, where $\lambda_1 = 0$, $\lambda_2, \ldots, \lambda_n$ are the eigenvalues of $L(\Gamma)$;

  (c) any cofactor of $L(\Gamma)$ (that is, the determinant of the matrix obtained by deleting row i and column j, multiplied by $(-1)^{i+j}$).

Example   If Γ is the complete graph on 3 vertices, then Γ has three spanning trees (each obtained by deleting one edge). We have

$$L(\Gamma) = \begin{pmatrix} 2 & -1 & -1 \\ -1 & 2 & -1 \\ -1 & -1 & 2 \end{pmatrix},$$

with eigenvalues $0, 3, 3$. A diagonal cofactor of $L(\Gamma)$ is

$$\begin{vmatrix} 2 & -1 \\ -1 & 2 \end{vmatrix} = 3,$$

while a typical off-diagonal cofactor is

$$-\begin{vmatrix} -1 & -1 \\ -1 & 2 \end{vmatrix} = 3.$$

Proof   The proof depends on the *Cauchy–Binet formula* , which says the following. Let A be an $m \times n$ matrix, and B an $n \times m$ matrix, where $m < n$. Then

$$\det(AB) = \sum_X \det(A(X)) \det(B(X)),$$

where X ranges over all m-element subsets of $\{1, \ldots, n\}$. Here $A(X)$ is the $m \times m$ matrix whose columns are the columns of A with index in X (as in the last section), and $B(X)$ is the $m \times m$ matrix whose rows are the rows of B with index in X.

Now choose an arbitrary orientation of the edges of Γ, and let M be the signed vertex-edge incidence matrix of Γ, as defined on page 164. It is straightforward to show that $MM^\top = L(\Gamma)$. Let v be any vertex of Γ, and let $N = M_v$ be the matrix obtained by deleting the row of M indexed by e. It can be shown that, if X is a set of $n-1$ edges, then

$$\det(N(X)) = \begin{cases} \pm 1 & \text{if } X \text{ is the edge set of a spanning tree,} \\ 0 & \text{otherwise.} \end{cases} \tag{9.2}$$

By the Cauchy–Binet formula, $\det(NN^\top)$ is equal to the number of spanning trees. But $NN^\top$ is the principal cofactor of $L(\Gamma)$ obtained by deleting the row and column indexed by v.

The fact that all cofactors are equal, and the expression in terms of eigenvalues, are proved by linear algebra; the proof is outlined in Exercise 9.6.

Example Let Γ be the complete graph on n vertices. Then $L(\Gamma)$ has diagonal entries $n-1$ and off-diagonal entries -1. Thus, $L(\Gamma) = nI - J$, where j denotes the all-1 matrix. We compute the eigenvalues of this matrix.

As we noted in general, the all-1 vector is an eigenvector of $L(\Gamma)$ with eigenvalue 0. We claim that the other $n-1$ eigenvalues are all equal to n. To see this, let $v = (v_1, \dots, v_n)$ be any vector orthogonal to the all-1 vector, that is, satisfying $v_1 + \cdots + v_n = 0$. Then $vJ = 0$, and so $v(nI + J) = nv$. So the claim is proved.

Now the formula in part (c) of the Matrix-Tree Theorem shows that the number of spanning trees is $(n^{n-1})/n = n^{n-2}$.

The fact that the number of trees on n vertices is n^{n-2} is *Cayley's Theorem*. We return to this theorem in the next chapter, where we will see several different proofs of it.

9.5   Exercises

9.1 Calculate the chromatic polynomials of the path and the cycle on n vertices.

9.2 Calculate the orbital chromatic polynomial $P_{\Gamma,G}(q)$, where

    (a) Γ is a path with n vertices and G consists of the identity and the reflection;

    (b) Γ is a cycle with n vertices and G is the cyclic group of order n consisting of rotations of the cycle.

9.3 Show that, if G is any group of automorphisms of a graph Γ, and q a positive integer, then $P_{\Gamma,G}(q) = 0$ if and only if $P_\Gamma(q) = 0$. Show that this is false if q is not assumed to be a positive integer.

9.4 Compute the Tutte polynomial of the complete graph K_4 on four vertices.

9.5   (a) Prove that, for an arbitrary graph Γ, the chromatic polynomial of Γ is equal to $q^c T(\Gamma; 1 - q, 0)$, where c is the number of connected components of Γ.

    (b) Modify the other parts of Proposition 9.4 so that they work for arbitrary graphs (not necesssarily connected).

9.6 Let A be a real symmetric matrix with all row and column sums zero. Let J denote the all-1 matrix. In this exercise we evaluate the determinant of $B = A + J$.

    (a) Replace the first row by the sum of all the rows; this makes the entries in the first row n and doesn't change the other entries; the determinant is unchanged.

    (b) Replace the first column by the sum of all the columns. This makes the first entry n^2, and the other entries in this column n, and doesn't change the other entries of the matrix; the determinant is unchanged.

(c) Subtract $1/n$ times the first row from each other row. The elements of the first column, other than the first, become 0; we subtract 1 from all elements not in the first row or column of B, leaving the entries of A; and the determinant is unchanged.

Conclude that $\det(B)$ is n^2 times the $(1,1)$ cofactor of A.

Check that this argument works for any cofactor of A.

Finally show that the all-1 vector is an eigenvector of B with eigenvalue n, while its other eigenvalues are the same as those of A. So $\det(B) = n\lambda_2 \cdots \lambda_n$.

9.7 Prove equation (9.2).

CHAPTER 10

Species

Species, invented by André Joyal in 1981, provide an attempt to unify some of the many structures and techniques which appear in combinatorial enumeration. I don't attempt to be too precise about what a species is. Think of it as a rule for building structures of some kind (graphs, posets, permutations, etc.) on any given finite set of points. We can ask for the number of labelled or unlabelled structures on n points in a given species.

10.1 Cayley's Theorem

We begin with a particular species where there is a simple but unexpected formula for the labelled counting problem. A *tree* is a connected graph with no cycles. It is straightforward to show that a tree on n vertices contains $n-1$ edges, and that any connected graph has a spanning tree (that is, some set of $n-1$ of its edges forms a tree). Moreover, any tree has a vertex lying on only one edge (since the average number of edges per vertex is $2(n-1)/n < 2$). Such a vertex is called a *leaf*. If we remove from a tree a leaf and its incident edge, the result is still a tree.

Cayley's Theorem states:

Theorem 10.1 *The number of labelled trees on n vertices is n^{n-2}.*

Remark A labelled tree on n vertices is just a spanning tree in the complete graph K_n. So, according to the last chapter, Cayley's Theorem is an evaluation of the Tutte polynomial of K_n. Moreover, in the last section, we deduced Cayley's Theorem from the Matrix-Tree Theorem.

There are many different proofs of Cayley's Theorem. Below, we will see two proofs which are made clearer by means of the concept of species. But first, one of the classics:

Prüfer's proof of Cayley's Theorem We construct a bijection between the set of all trees on the vertex set $\{1,\ldots,n\}$ and the set of all $(n-2)$-tuples of elements from this set. The tuple associated with a tree is called its *Prüfer code*.

First we describe the map from trees to Prüfer codes. Start with the empty code. Repeat the following procedure until only two vertices remain: select the leaf with smallest label; append the label of its unique neighbour to the code; and then remove the leaf and its incident edge.

Next, the construction of a tree from a Prüfer code P. We use an auxiliary list L of vertices added as leaves, which is initially empty. Now, while P is not empty, we join the first element of P to the smallest-numbered vertex v which is not in either P or L, and then add v to L and remove the first element of P. When P is empty, two vertices have not been put into L; the final edge of the tree joins these two vertices.

I leave to the reader the task of showing that these two constructions define inverse bijections. The method actually gives much more information:

Proposition 10.2 *In the tree with Prüfer code P, the valency of the vertex i is one more than the number of occurrences of i in P.*

For, at the conclusion of the second algorithm, if we add in the last two vertices to L, then L contains each vertex precisely once; and edges join each of the first $n-2$ vertices of L to the corresponding vertex in P, together with an edge joining the last two vertices of L.

Using this, one can count labelled trees with any prescribed degree sequence (see Exercise 10.10).

10.2 Species and counting

Almost the only thing we assume about a species $\mathcal{G}$ is that, for each n, there are only a finite number of $\mathcal{G}$-objects on a fixed base set of n points (so that we can count them). The only property we use of the objects in a species is that we 'know' whether a bijective map between the point sets of two objects is an isomorphism between them (and hence we know the automorphism group of each object).

We make one further (inessential but convenient) assumption, namely that there is a unique object on the empty set of points.

Joyal's view is that, just as a sequence of numbers can be represented by a formal power series, so a sequence of sets of mathematical objects can be represented by a 'combinatorial formal power series' (where the coefficient of x^n is the set of objects on the set $\{1,\ldots,n\}$). By replacing the set of objects by the number of labelled or unlabelled objects in the set, we obtain the usual generating functions for such objects.

We say that two species are *equivalent* (written $\mathcal{G} \sim \mathcal{H}$) if there is a bijection between the objects of the two species on a given point set such that the automorphism groups of corresponding objects are equal.

The most important formal power series associated with a species is its *cycle index*, which is defined by the rule

$$\tilde{Z}(\mathcal{G}) = \sum_{A \in \mathcal{G}} Z(\mathrm{Aut}(A)),$$

where $\mathrm{Aut}(A)$ is the automorphism group of A. Clearly, equivalent objects have the same cycle index.

The cycle index is well-defined since a monomial $s_1^{a_1} \cdots s_r^{a_r}$ arises only from cycle indices involving $n = \sum_{i=1}^{r} i a_i$ points, and by assumption there are only finitely many of these.

There are two important specialisations of the cycle index of a species $\mathcal{G}$; these are the exponential generating function

$$G(x) = \sum_{n \geq 0} \frac{G_n x^n}{n!}$$

for the number G_n of labelled n-element $\mathcal{G}$-objects (that is, objects on the point set $\{1,\ldots,n\}$); and the ordinary generating function

$$g(x) = \sum_{n \geq 0} g_n x^n$$

for the number g_n of unlabelled n-element $\mathcal{G}$-objects (that is, isomorphism classes).

Theorem 10.3 *Let $\mathcal{G}$ be a species. Then*

(a) $G(x) = \tilde{Z}(\mathcal{G}; s_1 \leftarrow x, s_i \leftarrow 0 \text{ for } i > 1)$;

(b) $g(x) = \tilde{Z}(\mathcal{G}; s_i \leftarrow x^i)$.

Proof The number of different labellings of an object A on n points is clearly $n!/|\mathrm{Aut}(A)|$. So it is enough to show that, for any permutation group G, we have

$$Z(G; s_1 \leftarrow x, s_i \leftarrow 0 \text{ for } i > 1) = x^n/|G|,$$
$$Z(G; s_i \leftarrow x^i) = x^n.$$

The first equation holds because putting $s_i = 0$ for all $i > 1$ kills all permutations except the identity. The second holds because, with this substitution, each group element contributes x^n, and the result is $1/|G| \sum_{g \in G} x^n = x^n$.

10.3 Examples of species

There are a few simple species for which we can do all the sums explicitly.

Example: Sets The species $\mathscr{S}$ has as its objects the finite sets, with one set of each cardinality up to isomorphism. Its cycle index was calculated in Chapter 7:

$$\tilde{Z}(\mathscr{S}) = \sum_{n \geq 0} (S_n) = \exp\left(\sum_{i \geq 1} \left(\frac{s_i}{i}\right)\right).$$

Hence we find that

$$\begin{aligned} S(x) &= \exp(x), \\ s(x) &= \exp\left(\sum_{i \geq 1} \frac{x^i}{i}\right) \\ &= \exp(-\log(1-x)) \\ &= \frac{1}{1-x}, \end{aligned}$$

in agreement with the fact that $S_n = s_n = 1$ for all $n \geq 0$.

Example: Total orders Let $\mathscr{L}$ be the species of total (or linear) orders. Each n-set can be totally ordered in $n!$ ways, all of which are isomorphic, and each of which is rigid (that is, has the trivial automorphism group).

We have

$$\tilde{Z}(\mathscr{L}) = \sum_{n \geq 0} s_1^n = \frac{1}{1 - s_1},$$

so that

$$L(x) = l(x) = \frac{1}{1-x}.$$

Example: Circular orders The species $\mathscr{C}$ consists of *circular orders*. An element of this species corresponds to placing the points of the object around a circle, where only the relative positions are considered, and there is no distinguished starting point. Thus, there is just one unlabelled n-element object in $\mathscr{C}$ for all n, and the number of labelled objects is equal to the number $(n-1)!$ of cyclic permutations for $n \geq 1$. The unique n-element structure has $\phi(m)$ automorphisms each with n/m cycles of length m for all m dividing n, where ϕ is Euler's function. Hence (see Exercise 7.2),

$$\tilde{Z}(\mathscr{C}) = 1 - \sum_{m \geq 1} \frac{\phi(m)}{m} \log(1 - s_m),$$

$$C(x) = 1 + \sum_{n \geq 1} \frac{x^n}{n} = 1 - \log(1-x),$$

$$c(x) = \sum x^n = \frac{1}{1-x}.$$

Example: Permutations An object of the species $\mathscr{P}$ consists of a set carrying a permutation. We will see later how $\mathscr{P}$ can be expressed as a composition, from which its cycle index can be deduced (Exercise 7.2). We have

$$\tilde{Z}(\mathscr{P}) = \prod_{n \geq 1} (1 - s_n)^{-1},$$

$$P(x) = \frac{1}{1-x},$$

$$p(x) = \prod_{n \geq 1} (1 - x^n)^{-1}.$$

The function $p(x)$ is the generating function for number partitions. For, as we saw earlier, an unlabelled permutation is the same as a conjugacy class of permutations; and conjugacy classes are determined by their cycle structure.

10.4 Operations on species

There are several ways of building new species from old; only a few important ones are discussed here.

10.4.1 Products

Let $\mathscr{G}$ and $\mathscr{H}$ be species. We define the *product* $\mathscr{K} = \mathscr{G} \times \mathscr{H}$ as follows: an object of $\mathscr{K}$ on a set X consists of a distinguished subset Y of X, a $\mathscr{G}$-object on Y, and a $\mathscr{H}$-object on $X \setminus Y$.

Since these objects are chosen independently, it is easy to check that

$$\tilde{Z}(\mathscr{G} \times \mathscr{H}) = \tilde{Z}(\mathscr{G})\tilde{Z}(\mathscr{H}).$$

Since the generating functions for labelled and unlabelled structures are specialisations of the cycle index, we have similar multiplicative formulae for them.

For example, if $\mathscr{S}$, $\mathscr{G}$ and $\mathscr{G}^\circ$ are the species of sets, graphs, and graphs with no isolated vertices respectively, then

$$\mathscr{G} \sim \mathscr{S} \times \mathscr{G}^\circ.$$

10.4.2 Substitution

Let $\mathscr{G}$ and $\mathscr{H}$ be species. We define the *substitution* $\mathscr{K} = \mathscr{G}[\mathscr{H}]$ as follows: an object of $\mathscr{K}$ on a set X consists of a partition of X, an $\mathscr{H}$-object on each part of the partition, and a $\mathscr{G}$-object on the set of parts of the partition.

Alternatively, we may regard it as a $\mathscr{G}$-object in which every point is replaced by a *non-empty $\mathscr{H}$*-object.

The cycle index is obtained from that of $\mathscr{G}$ by the substitution

$$s_i \leftarrow \tilde{Z}(\mathscr{H}; s_j \leftarrow s_{ij}) - 1$$

for all i. (The -1 in the formula corresponds to removing the empty $\mathscr{H}$-structure before substituting.)

From this, we see that the exponential generating functions for labelled structures obey the simple substitution law:

$$K(x) = G(H(x) - 1).$$

The situation for unlabelled structures is more complicated, and $k(x)$ cannot be obtained from $g(x)$ and $h(x)$ alone. Instead, we have

$$k(x) = \tilde{Z}(\mathscr{G}; s_i \leftarrow h(x^i) - 1).$$

This equation also follows from the Cycle Index Theorem, since we are counting functions on $\mathscr{G}$-structures where the figures are non-empty $\mathscr{H}$-structures with weight equal to cardinality.

For example, if $\mathscr{S}$, $\mathscr{P}$ and $\mathscr{C}$ are the species of sets permutations, and circular orders, then the standard decomposition of a permutation into disjoint cycles can be written

$$\mathscr{P} \sim \mathscr{S}[\mathscr{C}].$$

The counting series for labelled structures are given by

$$
\begin{aligned}
S(x) &= \sum_{n \geq 0} \frac{x^n}{n!} = \exp(x), \\
P(x) &= \sum_{n \geq 0} \frac{n! x^n}{n!} = \frac{1}{1-x}, \\
C(x) &= 1 + \sum_{n \geq 0} \frac{(n-1)! x^n}{n!} = 1 - \log(1-x);
\end{aligned}
$$

so the equation above becomes

$$\frac{1}{1-x} = \exp(-\log(1-x)).$$

So the decomposition of a permutation into cycles is the combinatorial equivalent of the fact that exp and log are inverse functions!

More generally, for any species $\mathscr{G}$, if $\mathscr{H} = \mathscr{S}[\mathscr{G}]$, then $H(x) = \exp(G(x))$. A typical situation for this is when every structure in $\mathscr{H}$ can be written in a unique way as a disjoint union of structures in $\mathscr{G}$, that is, $\mathscr{G}$ consists of the indecomposable structures in $\mathscr{H}$ and there is a nice decomposition theorem.

For example, this situation occurs if $\mathscr{G}$ is the species of connected graphs and $\mathscr{H}$ the species of all graphs.

This is the *exponential principle* that we met in Chapter 4. For an abstract treatment of the principle, see the paper of Dress and Müller in the bibliography.

10.4.3 Rooted (or pointed) structures

Given a species $\mathscr{G}$, let $\mathscr{G}^*$ be the species of *rooted $\mathscr{G}$-structures*: such a structure consists of a $\mathscr{G}$-structure with a distinguished point.

We have

$$\tilde{Z}(\mathscr{G}^*) = s_1 \frac{\partial}{\partial s_1} \tilde{Z}(\mathscr{G}),$$

and so

$$G^*(x) = x \frac{\mathrm{d}}{\mathrm{d}x} G(x).$$

Sometimes it is convenient to remove the distinguished point. This just removes the factors s_1 and x in the above formulae, so that this operation corresponds to differentiation. As a result, we denote the result by $\mathscr{G}'$.

For example, if $\mathscr{C}$ is the class of cycles, then $\mathscr{C}'$ corresponds to the class $\mathscr{L}$ of total (linear) orders. We have

$$L(x) = \frac{\mathrm{d}}{\mathrm{d}x} C(x) = \frac{\mathrm{d}}{\mathrm{d}x}(1 - \log(1 - x)) = \frac{1}{1-x},$$

in agreement with the preceding example.

10.5 Cayley's Theorem revisited

The notion of species can be used to give two further proofs of Cayley's Theorem.

First proof Let $\mathscr{L}$ and $\mathscr{P}$ be the species of total (or linear) orders and permutations, respectively. These species are quite different, but have the property that the numbers of labelled objects on n points are the same (namely $n!$).

Hence the numbers of labelled objects in the two species $\mathscr{L}[\mathscr{T}^*]$ and $\mathscr{P}[\mathscr{T}^*]$ are equal. (Here $\mathscr{T}^*$ is the species of rooted trees.)

Consider an object in $\mathscr{L}[\mathscr{T}^*]$. This consists of a linear order $(x_1, \ldots, x_r)$, with a rooted tree T_i at x_i for all i. I claim that this is equivalent to a tree with two distinguished vertices. Take edges $\{x_i, x_{i+1}\}$ for $i = 1, \ldots, r-1$, and identify x_i with the root of T_i for all i. The resulting graph is a tree. Conversely, given a tree with two distinguished vertices x and y, there is a unique path from x to y in the tree, and the remainder of the tree consists of rooted trees attached to the vertices of the path.

Now consider an object in $\mathscr{P}[\mathscr{T}^*]$. Identify the root of each tree with a point of the set on which the permutation acts, and orient each edge of this tree towards the root. The resulting structure defines a function f on the point set, where

- if v is a root, then $f(v)$ is the image of v under the permutation;

- if v is not a root, then $f(v)$ is the unique vertex for which $(v, f(v))$ is a directed edge of one of the trees.

Conversely, given a function $f : X \to X$, the set Y of periodic points of f has the property that f induces a permutation on it; the pairs $(v, f(v))$ for which v is not a periodic point have the structure of a family of rooted trees, attached to Y at the point for which the iterated images of v under f first enter Y.

So the number of trees with two distinguished points is equal to the number of functions from the vertex set to itself. Thus, if there are $F(n)$ labelled trees, we see that

$$n^2 F(n) = n^n,$$

from which Cayley's Theorem follows.

Second proof   As in the preceding proof, let $\mathscr{T}^*$ denote the species of rooted trees. If we remove the root from a rooted tree, the result consists of an unordered collection of trees, each of which has a natural root (at the neighbour of the root of the original tree). Conversely, given a collection of rooted trees, add a new root, joined to the roots of all the trees in the collection, to obtain a single rooted tree. So, if $\mathscr{E}$ denotes the species consisting of a single 1-vertex structure, and $\mathscr{S}$ the species of sets, we have

$$\mathscr{T}^* \sim \mathscr{E} \times \mathscr{S}[\mathscr{T}^*].$$

Hence, for the exponential generating functions for labelled structures, we have

$$T^*(x) = x \exp(T^*(x)).$$

This is, formally, a recurrence relation for the coefficients of $T^*(x)$, and we need to show that the nth coefficient is n^{n-1}. This can be done most easily with the technique of *Lagrange inversion*, which is discussed in the next chapter.

10.6   What is a species?

We have proceeded this far without ever giving a precise definition of a species. The informal idea is that an object of a species is constructed from a finite set, and bijections between finite sets induce isomorphisms of the objects built on them.

It turns out that mathematics does provide a language to describe this, namely *category theory*. It would take us too far afield to give all the definitions here. In essence, a category consists of a collection of *objects* with a collection of *morphisms* between them. In the only case with which we deal, objects are sets and morphisms are set mappings. In particular, the class $\mathfrak{S}$ whose objects are all finite sets and whose morphisms are all bijections between them satisfies the axioms for a category.

Now a species is simply a *functor* F from $\mathfrak{S}$ to itself. This means that F associates to each finite set S a set $F(S)$, and to each bijection $f : S \to S'$ a bijection $F(f) : F(S) \to F(S')$, such that F respects composition and identity (that is, $F(f_1 f_2) = F(f_1)F(f_2)$ and $F(1_S) = 1_{F(S)}$, where 1_S is the identity map on S).

The standard reference on species (apart from Joyal's original paper) is the book by Bergeron, Labelle and Leroux.

10.7 Oligomorphic permutation groups

Species are extremely general; while there are enumeration results about particular species or constructions, it is unlikely that there will be general results for all species. The species to be considered in this section have considerably stronger properties, and yet are general enough to include a wide range of examples. Also, the counting problems are clearly defined; there is no need to wonder what is the correct definition of the objects to be counted. For more details, see a recent survey in the ISI's *Prospects in Mathematical Sciences* series, given in the bibliography.

Let G be a permutation group on a set X, usually assumed infinite. There are several related actions of G: on X^n (the set of all n-tuples of elements of X, with coordinatewise action); on the set of n-tuples of distinct elements of X; and on the set of n-element subsets of X. We say that the group G is *oligomorphic* if the number of orbits in each of these actions is finite, for every natural number n. Indeed, for fixed n, finiteness of the number of orbits in one of these actions implies finiteness in each of the others, and also on m-tuples or m-sets for every $m < n$.

Now we have something to count: let $F_n^*(G)$ be the number of orbits of G on X^n; let $F_n(G)$ be the number of orbits on ordered n-tuples of distinct elements; and let $f_n(G)$ be the number of orbits on n-element subsets of X. Now the statements about finiteness can be quantified:

Proposition 10.4 *Let G be an oligomorphic permutation group.*

(a) $F_n^*(G) = \sum_{k=1}^{n} S(n,k)F_k(G)$ *for $n > 0$, where $S(n,k)$ is the Stirling number of the second kind.*

(b) $f_n(G) \leq F_n(G) \leq n! \, f_n(G)$.

(c) $F_n(G) \leq F_{n+1}(G)$, *with equality if and only if* $F_{n+1}(G) = 1$.

(d) $f_n(G) \leq f_{n+1}(G)$.

What does this have to do with combinatorics? Given a permutation group G on X, each orbit of G on X^n can be regarded as an n-ary relation on X which is invariant under G. Thus, G is a group of automorphisms of a relational structure. The finite induced substructures of this structure can be regarded as a species, if we so desire. (The term *relational structure* simply means a set carrying a collection of relations of various arities.) The *age* of a relational structure M is the collection of finite structures which are isomorphic to substructures of M.

The crucial definition is as follows. Let M be a relational structure on a set X. We say that M is *homogeneous* if the following holds: for every isomorphism $f : A \to B$ between finite induced substructures of M, there is an automorphism of M whose restriction to A is f. Now it can be shown that, for any permutation group G on X, the relational structure whose relations are all the orbits of G on n-tuples is homogeneous. In the other direction, the following holds:

Proposition 10.5 *Let M be a homogeneous relational structure, and G its automorphism group. Then*

(a) $F_n(G)$ *is equal to the number of labelled n-element structures in the age of M;*

(b) $f_n(G)$ *is equal to the number of unlabelled n-element structures in the age of M.*

This result is supplemented by a remarkable theorem of Fraïssé, which tells us how to recognise the age of a homogeneous relational structure. A class $\mathscr{C}$ of finite structures has the *amalgamation property* if, whenever $A, B_1, B_2 \in \mathscr{C}$ and $f_1 : A \to B_1$ and $f_2 : A \to B_2$ are embeddings (as induced substructures), there is a structure $C \in \mathscr{C}$ and embeddings $g_1 : B_1 \to C$ and $g_2 : B_2 \to C$ such that $f_1 g_1 = f_2 g_2$ (with composition from left to right). In other words, if two members of $\mathscr{C}$ have isomorphic substructures, they can be amalgamated or 'glued together' according to this isomorphism inside a larger structure in $\mathscr{C}$.

Theorem 10.6 *Let $\mathscr{C}$ be a class of finite relational structures. Suppose that the following conditions hold:*

(a) $\mathscr{C}$ *is closed under isomorphism;*

(b) $\mathscr{C}$ *is closed under taking induced substructures;*

(c) $\mathscr{C}$ *contains only countably many members up to isomorphism;*

(d) $\mathscr{C}$ *has the amalgamation property.*

Then there is a countable homogeneous structure M whose age is $\mathscr{C}$. Moreover, M is unique up to isomorphism with these properties. Conversely, the age of a countable homogeneous relational structure satisfies (a)–(d).

So now we know exactly which species arise in this way, and which have enumeration problems equivalent to orbit-counting problems for oligomorphic groups: they are precisely the *Fraïssé classes*, those which satisfy the conditions (a)–(d) of Fraïssé's Theorem.

Many classes satisfy these conditions: among them are the classes of finite partitions, graphs, directed graphs, tournaments, partially ordered sets, and so on. For a slightly different example, consider finite sets which are partitioned into sets of size at most 2, the parts being totally ordered. This can be shown to be a Fraïssé class. An n-element structure is specified up to isomorphism by an ordered sequence of 1s and 2s with sum n; so the number of n-element unlabelled structures in the class is the Fibonacci number F_n.

Investigations on enumeration of Fraïssé classes (that is, counting orbits of oligomorphic permutation groups) suggest that the counting sequences for such classes should grow *rapidly* and *smoothly*. There are some general results pointing in this direction. Here is just one such result, due to Francesca Merola (improving a result of Dugald Macpherson). A permutation group G on a set X is said to be *primitive* if there is no non-trivial partition of X which is invariant under G. (The trivial partitions are the partition with a single part and the partition into singletons.)

Theorem 10.7 *There is a universal constant $c > 1$ (in fact, $c > 1.324$) with the following property. Let G be a primtive oligomorphic permutation group, and assume that $f_n(G) > 1$ for some n. Then $f_n(G) \geq c^n/p(n)$ and $F_n(G) \geq n!\,c^n/q(n)$, where p and q are polynomials depending on G.*

It is conjectured that the 'correct' constant c in the theorem is $c = 2$.

There is far more to the rich theory of oligomorphic permutation groups. I mention here briefly some of the highlights, referring to the article cited in the bibliography for more details.

- It is possible to define a version of the cycle index for an oligomorphic permutation group G. Both the exponential generating function of the sequence $(F_n(G))$ and the ordinary generating function of the sequence $(f_n(G))$ arise as specialisations of this cycle index.

- There is a graded algebra A^G associated with the oligomorphic group G, with the property that the dimension of the nth homogeneous component of A^G is $f_n(G)$ (so that the *Hilbert series* of the algebra is equal to the ordinary generating function of the sequence $(f_n(G))$). Maurice Pouzet recently proved a long-standing conjecture asserting that, if G has no finite orbits, then A^G is an integral domain. This has implications for the smoothness of the sequence $(f_n(G))$: in particular, $f_{m+n}(G) \geq f_m(G) + f_n(G) - 1$ for all m, n.

- There are group-theoretic constructions (direct product, wreath product, and point stabiliser) which mirror the operations of product, substitution, and rooting of species.

- A remarkable theorem of first-order logic, proved independently by Engeler, Ryll-Nardzewski and Svenonius in 1959, asserts that a countable structure over a first-order language is *countably categorical* (that is, determined up to isomorphism by its first-order theory and the assumption of countability) if and only if its automorphism group is oligomorphic. For such structures M, the sequence $(F_n^*(\mathrm{Aut}(M)))$ counts the number of n-types over the theory of M. This remarkable result asserts the equivalence of *axiomatisability* (the structure is uniquely determined by first-order properties together with countability) and *symmetry* (the oligomorphic property of the automorphism group is a very strong symmetry condition).

Remark The prototype of a structure to which the theorem of Engeler *et al.* applies is the ordered set $(\mathbb{Q}, <)$ of rational numbers. A famous theorem of Cantor asserts that $\mathbb{Q}$ is, up to isomorphism, the unique countable totally ordered set which is

- *dense*, that is, given $x < y$, we can find z with $x < z < y$; and

- *without least or greatest element*, that is, for any x we can find y and z with $y < x$ and $x < z$.

The definition of a total order, together with the two conditions in the above bullet points, are first-order; so $(\mathbb{Q}, <)$ is countably categorical. You are invited to prove that its automorphism group is oligomorphic.

10.8 Weights

The theory of species allows us to interpret the statement

'Every graph is uniquely expressible as the disjoint union of connected graphs'

as the relation $\mathscr{G} \sim \mathscr{S}[\mathscr{C}]$, where $\mathscr{S}$, $\mathscr{G}$ and $\mathscr{C}$ are the species of sets, graphs and connected graphs respectively.

We can get extra information if the objects in our species carry *weights*. The weights must satisfy some restrictions to make the process work nicely. Rather than describe these in general, I consider one special case. We give each edge a weight (which, in general, may differ from edge to edge, but in the special case here will be the same for each edge). Now the weight of a graph, or connected graph, is just the product of its edge weights; and so the weight of a graph is the product of the weights of its connected components. This means that

$$\mathcal{G}_w \sim \mathcal{S}[\mathcal{C}_w],$$

where $\mathcal{G}_w$ and $\mathcal{C}_w$ are the species of weighted graphs and weighted connected graphs respectively.

Now specialise further, to the case where $w = -1$, and consider the exponential generating function for labelled objects. For graphs on n vertices, if $n > 1$, the generating function is

$$\sum_{E \subseteq \binom{\{1,\ldots,n\}}{2}} (-1)^{|E|} = \sum_{i=0}^{\binom{n}{2}} (-1)^i \binom{\binom{n}{2}}{i} = (1-1)^{\binom{n}{2}} = 0.$$

So $G_w(x) = 1 + x$. If $c_{n,k}$ is the number of connected labelled graphs with n vertices and k edges, then

$$C_w(x) = \sum_n \sum_{k=0}^{\binom{n}{2}} \frac{(-1)^k c_{n,k}}{n!} x^n.$$

And as we saw earlier, $S(x) = \exp(x)$.

So we have

$$1 + x = \exp \sum_n \left(\sum_{k=0}^{\binom{n}{2}} \frac{(-1)^k c_{n,k}}{n!} \right) x^n,$$

or in other words,

$$\sum_n \left(\sum_{k=0}^{\binom{n}{2}} \frac{(-1)^k c_{n,k}}{n!} \right) x^n = \log(1+x) = \sum_n \frac{(-1)^{n-1}}{n} x^n.$$

Equating coefficients, we find the pretty formula for the alternating sum of the numbers of connected graphs:

$$\sum_{k=0}^{\binom{n}{2}} (-1)^k c_{n,k} = (-1)^{n-1} (n-1)!.$$

(The lower limit in the sum can be taken to be $n - 1$, since this is the smallest number of edges in a connected graph.)

Note that the terms in the alternating sum are much larger than the final value.

Example For $n = 4$, there are $4^2 = 16$ trees (connected graphs with 3 edges), by Cayley's Theorem. Any graph with at least four edges is necessarily connected, so the numbers of graphs are $\binom{6}{4} = 15$, $\binom{6}{5} = 6$ and $\binom{6}{6} = 1$ for 4, 5 and 6 edges. So the alternating sum in the theorem is $-16 + 15 - 6 + 1 = -6 = -3!$.

This argument was taken from a talk by Stephen Tate, who also had a formula for the alternating sum of numbers of 2-connected graphs (*blocks*), based on a slightly more complicated species equation.

The use of weights in this way is important in the connections between combinatorics and statistical mechanics, where typically the vertices of a graph represent particles, and edge weights carry information about the strengths of their interactions.

10.9   Exercises

10.1 Count the labelled trees in which the vertex i has valency a_i for $1 \le i \le n$, where $a_1, \ldots, a_n$ are positive integers with sum $2n - 2$.

10.2 Let π be a permutation on $\{1, \ldots, n\}$ consisting of a single n-cycle. Show that the minimum number of transpositions whose product is π is $n - 1$, and that a product of $n - 1$ transpositions is an n-cycle if and only if the cycles of the transpositions are the edges of a tree. Hence show that the number of ways of writing π as a product of $n - 1$ transpositions is n^{n-2}. (Hint: count pairs consisting of a tree with the edges ordered and an n-cycle, where the product of the transpositions corresponding to edges of the tree, in the given order, is the n-cycle.)

10.3 Show that the cycle index for the species $\mathscr{C}$ of circular structures is

$$\tilde{Z}(\mathscr{C}) = 1 - \sum_{m \ge 1} \frac{\phi(m)}{m} \log(1 - s_m).$$

Use the fact that

$$\mathscr{P} \sim \mathscr{S}[\mathscr{C}]$$

to show that

$$\tilde{Z}(\mathscr{P}) = \prod_{n \ge 1} (1 - s_n)^{-1}.$$

Can you give a direct proof of this?

10.4 Use the result of the preceding exercise, and the fact that $c_n = 1$ for all n (where c_n is the number of unlabelled n-element structures in $\mathscr{C}$) to prove the identity

$$\prod_{m \ge 1} (1 - x^m)^{-\phi(m)/m} = \exp(x/(1 - x)).$$

10.5 Suppose that g_n is the number of unlabelled n-element objects in the species $\mathcal{G}$. Show that the generating function for unlabelled structures in $\mathcal{S}[\mathcal{G}]$ is

$$\prod_{n\geq 1}(1-x^n)^{-g_n}.$$

Verify this combinatorially in the case $\mathcal{G} = \mathcal{S}$. How would you describe the objects of $\mathcal{S}[\mathcal{S}]$?

10.6 Let $\mathcal{S}$, $\mathcal{P}$ and $\mathcal{D}$ be the species of sets, permutations and derangements respectively. By 'decomposing' a permutation as a set of fixed points and a derangement of the remaining points, show that

$$\mathcal{P} \sim \mathcal{S} \times \mathcal{D}.$$

Use this to deduce the exponential generating function for the number of derangements of an n-set.

10.7 Let $\mathcal{G}$ be a species. The *Stirling numbers* of $\mathcal{G}$ are the numbers $S(\mathcal{G})(n,k)$, defined to be the number of partitions of an n-set into k parts with a $\mathcal{G}$-object on each part.

(a) Prove that, for $\mathcal{G} = \mathcal{S}, \mathcal{C}$ and $\mathcal{L}$ respectively, the Stirling numbers are respectively the Stirling numbers $S(n,k)$ of the second kind, the unsigned Stirling numbers $|s(n,k)|$ of the first kind, and the Lah numbers $L(n,k)$ respectively.

(b) Let $S(\mathcal{G})$ be the lower triangular matrix of Stirling numbers of $\mathcal{G}$. Prove that

$$S(\mathcal{G})S(\mathcal{H}) = S(\mathcal{H}[\mathcal{G}]).$$

(c) Let (a_n) and (b_n) be sequences of positive integers with exponential generating functions $A(x)$ and $B(x)$ respectively. Prove that the following two conditions are equivalent:

- $a_0 = b_0$ and $b_n = \displaystyle\sum_{k=1}^{n} S(\mathcal{G})(n,k)a_k$ for $n \geq 1$;
- $B(x) = A(G(x) - 1)$.

10.8 A *forest* is a graph whose connected components are trees. Show that there is a bijection between labelled forests of rooted trees on n vertices, and labelled rooted trees on $n+1$ vertices with root $n+1$.

Hence show that, if a forest of rooted trees on n vertices is chosen at random, then the probability that it is connected tends to the limit $1/e$ as $n \to \infty$.

Remark It is true but harder to prove that the analogous limit for unrooted trees is $1/\sqrt{e}$.

10.9 Let $\mathcal{U}$ be the *subset* species: a $\mathcal{U}$-object consists of a distinguished subset of its ground set. Calculate the cycle index of $\mathcal{U}$. Hence or otherwise prove that the enumeration functions of $\mathcal{U}$ are

$$
\begin{aligned}
U(x) &= \exp(2x), \\
u(x) &= (1-x)^{-2}.
\end{aligned}
$$

10.10 Count the labelled trees in which the vertex i has valency a_i for $1 \le i \le n$, where $a_1, \ldots, a_n$ are positive integers with sum $2n - 2$.

10.11 A *forest* is a graph whose connected components are trees. Show that there is a bijection between labelled forests of rooted trees on n vertices, and labelled rooted trees on $n+1$ vertices with root $n+1$.

Hence show that, if a forest of rooted trees on n vertices is chosen at random, then the probability that it is connected tends to the limit $1/e$ as $n \to \infty$.

10.12 Let $\mathcal{L}$ and $\mathcal{E}$ denote the species of linear orders and of sets of size at most 2 respectively. Prove that the number of labelled n-element structures in $\mathcal{L}[\mathcal{E}]$ is the Fibonacci number F_n.

10.13 (a) Prove the first three parts of Proposition 10.4. (The fourth part is more difficult.)

  (b) Prove Proposition 10.5.

  (c) Prove that the relational structure on X whose relations are the orbits on n-tuples of a permutation group G on X is homogeneous.

10.14 (a) Prove that the class of finite graphs is a Fraïssé class.

  (b) Prove that the class of finite posets is a Fraïssé class.

  (c) Prove that the class of structures consisting of a finite set partitioned into parts of size at most 2 with the set of parts totally ordered (mentioned in the text in connection with Fibonacci numbers) is a Fraïssé class.

10.15 Let $(a_1, \ldots, a_n)$ and $(b_1, \ldots, b_n)$ be two n-tuples of rational numbers, both in strictly increasing order. Show that there exists an order-preserving permutation of $\mathbb{Q}$ carrying a_i to b_i for $i = 1, \ldots, n$. Deduce that, for $G = \mathrm{Aut}(\mathbb{Q}, <)$, we have

$$
f_n(G) = 1, \qquad F_n(G) = n!, \qquad F_n^*(G) = \sum_{k=1}^{n} S(n,k) k!.
$$

10.16   (a) Let G and H be oligomorphic permutation groups on sets X and Y respectively. There is a natural action of the direct product $G \times H$ on the Cartesian product $X \times Y$, given by the rule

$$(x, y)^{(g,h)} = (x^g, y^h).$$

Prove that $G \times H$, with this action, is oligomorphic, and that

$$F_n^*(G \times H) = F_n^*(G) F_n^*(H).$$

(b) Now let G be the group of order-preserving permutations of $\mathbb{Q}$, as defined in the preceding question. Write down a formula for $F_n(G \times G)$ (with the product action defined in the first part of this question). Show also that $f_n(G \times G) = F_n(G \times G)/n!$.

(c) Show that $f_n(G \times G)$ is equal to the number of matrices (of any size) of zeros and ones having precisely n ones and having no row or column consisting entirely of zeros.

(This result is due to Cameron, Gewurz and Merola; the asymptotics of the numbers $f_n(G)$ were calculated by Cameron, Prellberg and Stark.)

CHAPTER 11

Analytic methods: a first look

In this chapter, we say a bit more about asymptotics, prove Stirling's formula, consider the use of complex analysis in combinatorics, and finally describe subadditive and submultiplicative functions, for which the shape of the asymptotics is easy to establish, although precise details may be more difficult.

11.1   The language of asymptotics

We introduce the notation for describing the asymptotic behaviour of functions here.

Let F and G be functions of the natural number n. For convenience we assume that G does not vanish. We write

- $F = O(G)$ if $F(n)/G(n)$ is bounded above as $n \to \infty$;
- $F = \Omega(G)$ if $F(n)/G(n)$ is bounded below as $n \to \infty$;
- $F = \Theta(G)$ if $F(n)/G(n)$ is bounded above and below as $n \to \infty$;
- $F = o(G)$ if $F(n)/G(n) \to 0$ as $n \to \infty$;
- $F \sim G$ if $F/G \to 1$ as $n \to \infty$.

If $F \sim G$, then the difference satisfies $F - G = o(G)$. This might be accompanied by an asymptotic estimate for $F(n) - G(n)$, and so on; we obtain an *asymptotic series* for F.

More precisely, the series $G_0(n) + G_1(n) + G_2(n) + \cdots$ is an *asymptotic series* for $F(n)$ if

$$F(n) - \sum_{j=0}^{i-1} G_j(n) \sim G_i(n)$$

for $i \geq 0$. (So in particular $F(n) \sim G_0(n)$, $F(n) - G_0(n) \sim G_1(n)$, and so on. Note that $G_i(n) = o(G_{i-1}(n))$ for all i.)

The definition comes with a couple of warnings:

- an asymptotic series is not necessarily convergent;

- it is not necessarily the case that taking more terms in the series gives a better approximation to $F(n)$ for a fixed n.

Typically, F is a combinatorial enumeration function, and G a combination of standard functions of analysis. In the next section, we prove *Stirling's formula* as an example.

11.2 Stirling's formula

Stirling's formula gives the asymptotics of the number $n!$ of permutations of $\{1, \ldots, n\}$. We give the proof as an illustration.

Theorem 11.1

$$n! \sim \sqrt{2\pi n} \left(\frac{n}{e} \right)^n.$$

Proof Consider the graph of the function $y = \log x$ between $x = 1$ and $x = n$, together with the piecewise linear functions shown in Figure 11.1.

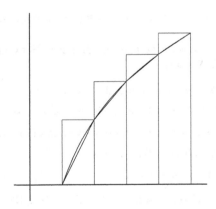

Figure 11.1: Stirling's formula

Let $f(x) = \log x$, let $g(x)$ be the function whose value is $\log m$ for $m \leq x < m+1$, and let $h(x)$ be the function defined by the polygon with vertices $(m, \log m)$, for

$1 \leq m \leq n$. Clearly

$$\int_1^n g(x)\,dx = \log 2 + \cdots + \log n = \log n!\,.$$

The difference between the integrals of g and h is the sum of the areas of triangles with base 1 and total height $\log n$; that is, $\frac{1}{2}\log n$.

Some calculus (given at the end of this proof) shows that the difference between the integrals of f and h tends to a finite limit c as $n \to \infty$.

Finally, a simple integration shows that

$$\int_1^n f(x)\,dx = n\log n - n + 1.$$

We conclude that

$$\log n! = n\log n - n + \tfrac{1}{2}\log n + (1 - c) + o(1),$$

so that

$$n! \sim \frac{Cn^{n+1/2}}{e^n}.$$

To identify the constant C, we can proceed as follows. Consider the integral

$$I_n = \int_0^{\pi/2} \sin^n x\,dx.$$

Integration by parts shows that

$$I_n = \frac{n-1}{n}I_{n-2},$$

and hence

$$I_{2n} = \frac{(2n)!\,\pi}{2^{2n+1}(n!)^2},$$

$$I_{2n+1} = \frac{2^{2n}(n!)^2}{(2n+1)!}.$$

On the other hand,

$$I_{2n+2} \leq I_{2n+1} \leq I_{2n},$$

from which we get

$$\frac{(2n+1)\pi}{4(n+1)} \leq \frac{2^{4n}(n!)^4}{(2n)!(2n+1)!} \leq \frac{\pi}{2},$$

and so

$$\lim_{n\to\infty} \frac{2^{4n}(n!)^4}{(2n)!(2n+1)!} = \frac{\pi}{2}.$$

Putting $n! \sim Cn^{n+1/2}/e^n$ in this result, we find that

$$\frac{C^2 e}{4} \lim_{n\to\infty} \left(1 + \frac{1}{2n}\right)^{-(2n+3/2)} = \frac{\pi}{2},$$

so that $C = \sqrt{2\pi}$.

The last part of this proof is taken from Alan Slomson's *An Introduction to Combinatorics* (see the bibliography). It is more-or-less the same as the proof of Wallis' product formula for π.

Here for completeness is the argument we skipped over in the proof.

Let $F(x) = f(x) - h(x)$. The convexity of $\log x$ shows that $F(x) \geq 0$ for all $x \in [m, m+1]$. For an upper bound we use the fact, a consequence of Taylor's Theorem, that

$$\log x \leq \log m + \frac{x-m}{m} \leq \log m + \frac{1}{m}$$

for $x \in [m, m+1]$. Then

$$F(x) = \log x - \log m - \log\left(1 + \frac{1}{m}\right)(x - m) \leq \frac{1}{m} - \log\left(1 + \frac{1}{m}\right) \leq \frac{1}{2m^2},$$

where the last inequality comes from another application of Taylor's Theorem which yields $\log(1 + x) \geq x - x^2/2$ for $x \in [0, 1]$. Now $\sum(1/m^2)$ converges, so the integral is bounded.

11.3   Complex analysis

Regarding a generating function for a sequence as a function of a real or complex variable is a powerful method for studying the asymptotic behaviour of the sequence. We will see examples of this later. The simplest applications are based on the following fundamental theorem about the behaviour of power series in complex analysis. Here, we reiterate the application of complex power series to asymptotics, and give an example, finding the asymptotic number of partial preorders.

Recall that, if x_n is a sequence of real numbers, then the *limit superior* (for short, lim sup) of the sequence is the real number r (if it exists) with the property that, for any $\varepsilon > 0$,

- infinitely many terms of the sequence are greater than $r - \varepsilon$;
- only finitely many terms are greater than $r + \varepsilon$.

A sequence has a lim sup if and only if it is bounded. (Take r to be the infimum of the set of values y for which only finitely many terms x_n are greater than x. This set is non-empty (it contains any upper bound for the sequence) and bounded below (a lower bound for the sequence is also a lower bound for the set), and so the infimum exists; it clearly has the required property.)

There is an analogous dual notion of *limit inferior* or lim inf.

Theorem 11.2 *Suppose that $A(z) = \sum a_n z^n$ defines a function which is analytic in some neighbourhood of the origin in the complex plane. Suppose that the smallest modulus of a singularity of $A(z)$ is R. Then $\limsup |a_n|^{1/n} = 1/R$.*

This shows that all but finitely many a_n are bounded by $(c+\varepsilon)^n$, but infinitely many are not bounded by $(c-\varepsilon)^n$, where $c = 1/R$. We conclude that the series is 'roughly' c^n. (However, it is not true to say that $a_n \sim c^n$: why?)

On the other hand, if $A(z)$ is analytic everywhere, then $a_n \le \varepsilon^n$ for $n > n_0(\varepsilon)$, for any positive ε. Indeed, $a_n = o(\varepsilon^n)$ for any positive ε.

For example, if $B(n)$ is the nth Bell number, then

$$\sum_{n \ge 0} \frac{B(n)z^n}{n!} = e^{e^z - 1},$$

which is analytic everywhere. So $B(n) = o(\varepsilon^n n!)$, for any positive ε.

Here is an example. In Exercise 3.23, we defined a *total preorder* to be a reflexive and transitive relation on a set X which satisfies the *trichotomy law*, that is, for any $x, y \in X$, either $R(x,y)$ or $R(y,x)$ holds.

We showed that the number $P(n)$ of total preorders on the set $\{1, \ldots, n\}$ has exponential generating function given by

$$p(z) = \sum_{n \ge 0} \frac{P(n)z^n}{n!} = \frac{1}{2 - \exp(z)}.$$

We use this to find a very good asymptotic estimate for $P(n)$.

The generating function has a simple pole at $z = \log 2$ with residue

$$\lim_{z \to \log 2} \left(\frac{z - \log 2}{2 - e^z} \right) = \lim_{z \to \log 2} \left(\frac{1}{-e^z} \right) = -\frac{1}{2},$$

by l'Hôpital's rule. So the function

$$p(z) + \frac{1}{2(z - \log 2)}$$

is analytic in a circle with centre at the origin and the next singularities of $p(z)$ (at $\log 2 \pm 2\pi i$) on the boundary. Thus

$$P(n) \sim \frac{n!}{2} \left(\frac{1}{\log 2} \right)^{n+1},$$

and indeed the difference between the two sides is $n! \, o((r - \varepsilon)^{-n})$, where $r = |\log 2 + 2\pi i|$; that is, exponentially smaller than either expression.

For $n = 10$, the number of preorders is 102247563, while the asymptotic estimate is 102247563.005271042....

11.4   Subadditive and submultiplicative functions

This section describes a simple but quite powerful idea for estimating the asymptotics of a function on the natural numbers from a condition on its behaviour.

Let f be a function from the natural numbers to (say) the positive real numbers. We assume for simplicity that $f(0) = 0$. Then f is said to be *subadditive* if, for all positive integers m, n, we have

$$f(m+n) \le f(m) + f(n).$$

Since $f(0) = 0$, we can phrase this a different way:

$$\frac{f(m+n) - f(n)}{m} \le \frac{f(m) - f(0)}{m}.$$

This says that the slope of the secant from $(n, f(n))$ to $(m+n, f(m+n))$ is less than or equal to the slope of the secant from $(0,0)$ to $(m, f(m))$: see Figure 11.2.

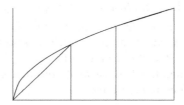

Figure 11.2: A subadditive function

Proposition 11.3 *Let f be a subadditive function. Then*

$$c = \lim_{n \to \infty} \frac{f(n)}{n}$$

exists; if $c > 0$ then we have $f(n) \sim cn$.

Proof   An easy induction shows that $f(m_1 + \cdots + m_s) \le f(m_1) + \cdots + f(m_s)$ for any $m_1, \ldots, m_s$. In particular, if $n = mq + r$, with $0 \le r < m$, then $f(n) \le qf(m) + f(r)$, and so

$$\frac{f(n)}{n} \le \frac{qf(m)}{mq+r} + \frac{f(r)}{mq+r} \le \frac{f(m)}{m} + \frac{P}{n},$$

where $P = \max\{f(0), \ldots, f(m-1)\}$.

Fixing m and letting $n \to \infty$, we see that

$$\limsup_{n \to \infty} \frac{f(n)}{n} \le \frac{f(m)}{m}.$$

for any value of m.

Choose a sequence $m_1, m_2, \ldots$ for which

$$\lim_{i \to \infty} \frac{f(m_i)}{m_i} = \liminf_{n \to \infty} \frac{f(n)}{n}.$$

We see that $\limsup f(n)/n \leq \liminf f(n)/n$, and so the limit exists, as claimed.

Note that the final conclusion requires that $c > 0$. The function $f(n) = n^a$ is subadditive for any positive real number $a < 1$; we have $\lim_{n \to \infty} f(n)/n = 0$, but the asymptotics of f are not determined by this fact.

We do not meet many subadditive functions in enumerative combinatorics. A much more common situation is that we have a function which is submultiplicative, in the following sense.

Let f be a function from the natural numbers to the positive reals; assume that $f(0) = 1$. We say that f is *submultiplicative* if, for all positive integers m, n we have

$$f(m+n) \leq f(m)f(n).$$

Proposition 11.4 *Let f be a submultiplicative function. Then*

$$c = \lim_{n \to \infty} f(n)^{1/n}$$

exists; if $c > 1$ then we have $\log f(n) \sim n \log c$.

Proof Apply Proposition 11.3 to the function $g(n) = \log f(n)$.

The final statement of the proposition asserts that $f(n)$ is 'roughly' c^n, that is, $f(n)$ grows exponentially with n; but it is not correct to conclude from our hypothesis that $f(n) \sim c^n$, as we will see.

Here is a very simple example. Let $f(n)$ be the number of strings of zeros and ones of length n containing no two consecutive ones. We claim that f is submultiplicative. For consider the strings of length $m+n$ satisfying the constraint. The first m positions form a string of length m with no two consecutive ones; there are $f(m)$ possibilities for this. The last n positions form a string of length n with no two consecutive ones; there are $f(n)$ possibilities. So there are $f(m)f(n)$ possible concatenations of such strings; not all pass the test, since we have to rule out the possibility that the first string ends with 1 while the second begins with 1. So certainly $f(m+n) \leq f(m)f(n)$.

In fact, as we saw in Chapter 2, we have $f(n) \sim Ac^n$, where $c = (1+\sqrt{5})/2$ and $A = (1+\sqrt{5})/2\sqrt{5}$.

The result obtained using submultiplicativity is of course much less precise, but much more general; it holds for the number of binary strings with an arbitrary finite collection of forbidden substrings, for example.

11.5   Exercises

11.1 Prove that $n^k = o(c^n)$ for any constants $k > 0$ and $c > 1$, and that $\log n = o(n^{\varepsilon})$ for any $\varepsilon > 0$.

11.2 Let R be a preorder on X.

(a) Define a relation $\equiv$ on X by the rule that $x \equiv y$ if $R(x,y)$ and $R(y,x)$ both hold. Prove that $\equiv$ is an equivalence relation.

(b) Define a relation $\leq$ on the set of equivalence classes of $\equiv$ by the rule that $[x] \leq [y]$ if $R(x,y)$ holds. Show that this relation is well-defined (independent of the choice of representatives of the equivalence classes) and is a total order.

(c) Show that, given any equivalence relation on X and any total order on the set of equivalence classes, there is a unique preorder on X which gives rise to them in the manner just described.

(Hence $P(n)$ is the number of ways of choosing a partition of the set $\{1,\dots,n\}$ and a total order of the set of parts.)

11.3 A *self-avoiding walk* of length n on the square lattice in the plane is a sequence of $n+1$ points $P_0,\dots,P_n$ with integer coordinates, such that

- $P_0 = (0,0)$;
- for $i = 1,\dots,n$, we have $P_i - P_{i-1} \in \{(1,0),(0,1),(-1,0),(0,-1)\}$;
- the points $P_0,\dots,P_n$ are all distinct.

In other words, start at the origin, take n steps each one unit east, north, west or south, such that no point is ever revisited.

Let $f(n)$ be the number of self-avoiding walks of order n. Prove that

(a) $4 \cdot 2^{n-1} \leq f(n) \leq 4 \cdot 3^{n-1}$;

(b) f is submultiplicative.

Deduce that $c = \lim_{n \to \infty} f(n)^{1/n}$ exists, and that $2 \leq c \leq 3$.

Remark The problem of actually finding the value of the limit c is extremely difficult!

11.4 Define *superadditive* and *supermultiplicative* functions by reversing the inequalities in the definition of subadditive and submultiplicative functions, and prove the analogues of Propositions 11.3 and 11.4.

CHAPTER 12

Further topics

In this section we consider some further topics, in rather less detail: Lagrange inversion, Bernoulli numbers, the Euler–Maclaurin sum formula, and analytic techniques due to Hayman, Meir and Moon, and Bender.

12.1 Lagrange inversion

A formal power series over a field, with zero constant term and non-zero term in x, has an inverse with respect to composition. Indeed, the set of all such formal power series is a group, which has recently become known as the *Nottingham group*. However, the basic facts are much older. The associative, closure, and identity laws are obvious, and the rule for finding the inverse in characteristic zero is known as *Lagrange inversion*.

12.1.1 The theorem

The basic fact can be stated as follows.

Proposition 12.1 *Let f be a formal power series over $\mathbb{R}$, with $f(0) = 0$ and $f'(0) \neq 0$. Then there is a unique formal power series g such that $g(f(x)) = x$; the coefficient of y^n in $g(y)$ is*

$$\left[\frac{d^{n-1}}{dx^{n-1}} \left(\frac{x}{f(x)} \right)^n \right]_{x=0} \Big/ n!.$$

This can be expressed in a more convenient way for our purpose. Let

$$\phi(x) = \frac{x}{f(x)}.$$

Then the inverse function g is given by the functional equation

$$g(y) = y\phi(g(y)).$$

Then Lagrange inversion has the form

$$g(y) = \sum_{n \geq 1} \frac{b_n y^n}{n!},$$

where

$$b_n = \left[\frac{d^{n-1}}{dx^{n-1}} \phi(x)^n \right]_{x=0}.$$

Example: Cayley's Theorem   The exponential generating function for rooted trees satisfies the equation

$$T^*(x) = x \exp(T^*(x)).$$

With $\phi(x) = \exp(x)$, we find that the coefficient of $y^n/n!$ in $T^*(y)$ is

$$\left[\frac{d^{n-1}}{dx^{n-1}} \exp(nx) \right]_{x=0} = n^{n-1},$$

proving Cayley's Theorem once again.

12.1.2   Proof of the theorem

The proof of Lagrange's inversion formula involves a considerable detour. The treatment here follows the book by Goulden and Jackson. Throughout this subsection, we assume that the coefficients form a field of characteristic zero; for convenience, we assume that the coefficient ring is $\mathbb{R}$.

First, we extend the notion of formal power series to *formal Laurent series*, defined to be a series of the form

$$f(x) = \sum_{n \geq m} a_n x^n,$$

where m may be positive or negative. If the series is not identically zero, we may assume without loss of generality that $a_m \neq 0$, in which case m is the *valuation* of f, written

$$m = \mathrm{val}(f).$$

We define addition, multiplication, composition, differentiation, etc., for formal Laurent series as for formal power series. In particular, $f(g(x))$ is defined for any formal Laurent series f, g with $\mathrm{val}(g) > 0$. (This is less trivial than the analogous result for formal power series.) In particular, we need to know that $g(x)^{-m}$ exists as a formal Laurent series for $m > 0$. It is enough to deal with the case $m = 1$, since certainly $g(x)^m$ exists. If $\mathrm{val}(g) = r$, then $g(x) = x^r g_1(x)$, and so $g(x)^{-1} = x^{-r}g_1(x)^{-1}$, and we have seen that $g_1(x)^{-1}$ exists as a formal power series, since $g_1(0)$ is invertible.

We denote the derivative of the formal Laurent series $f(x)$ by $f'(x)$.

We also introduce the following notation: $[x^n]f(x)$ denotes the coefficient of x^n in the formal power series (or formal Laurent series) $f(x)$. The case $n = -1$ is especially important, as we learn from complex analysis. The value of $[x^{-1}]f(x)$ is called the *residue* of $f(x)$, and is also written as $\mathrm{Res}\, f(x)$.

Everything below hinges on the following simple observation, which is too trivial to need a proof.

Proposition 12.2 *For any formal Laurent series $f(x)$, we have* $\mathrm{Res}\, f'(x) = 0$.

Now the following result describes the residue of the composition of two formal Laurent series.

Theorem 12.3 (Residue Composition Theorem) *Let $f(x)$, $g(x)$ be formal Laurent series satisfying* $\mathrm{val}(g) = r > 0$. *Then*

$$\mathrm{Res}(f(g(x))g'(x)) = r\,\mathrm{Res}(f(x)).$$

Proof It is enough to consider the case where $f(x) = x^n$, since Res is a linear function.

Suppose that $n \neq -1$, so that the right-hand side is zero. Then

$$\mathrm{Res}(g^n(x)g'(x)) = \frac{1}{n+1}\,\mathrm{Res}\left(\frac{\mathrm{d}}{\mathrm{d}x}g^{n+1}(x)\right) = 0.$$

So consider the case where $n = -1$. Let $g(x) = ax^r h(x)$, where $a \neq 0$ and $h(0) = 1$. Then

$$\begin{aligned}
g'(x) &= rax^{r-1}h(x) + ax^r h'(x), \\
\frac{g'(x)}{g(x)} &= \frac{r}{x} + \frac{h'(x)}{h(x)},
\end{aligned}$$

so

$$\mathrm{Res}\, g'(x)g(x)^{-1} = r = r\,\mathrm{Res}\, x^{-1},$$

since $h'(x)/h(x) = (\mathrm{d}/\mathrm{d}x)\log h(x)$, and $\log h(x) = \log(1 + k(x))$ is a formal power series since $k(x)$ is a f.p.s. with constant term zero.

It is tempting to say

$$g'(x)g(x)^{-1} = \frac{d}{dx}\log g(x)$$

$$= \frac{d}{dx}(\log a + r\log x + \log h(x))$$

$$= \frac{r}{x} + \frac{d}{dx}\log h(x),$$

but this is not valid; $\log g(x)$ may not exist as a formal Laurent series. Consider this point carefully; an error here would lead to the incorrect conclusion that $\mathrm{Res}(g'(x)/g(x)) = 0$.

From the Residue Composition Theorem, we can prove a more general version of Lagrange Inversion.

Theorem 12.4 (Lagrange Inversion) *Let ϕ be a formal power series satisfying* $\mathrm{val}(\phi) = 0$. *Then the equation*

$$g(x) = x\phi(g(x))$$

has a unique formal power solution $g(x)$. Moreover, for any Laurent series f, we have

$$[x^n]f(g(x)) = \begin{cases} \frac{1}{n}[x^{n-1}](f'(x)\phi(x)^n) & \text{if } n \geq \mathrm{val}(f) \text{ and } n \neq 0, \\ f(0) + \mathrm{Res}(f'(x)\log(\phi(0)^{-1}\phi(x))) & \text{if } n = 0. \end{cases}$$

Proof Let $\Phi(x) = x/\phi(x)$, so that $\Phi(g(x)) = x$ and $\mathrm{val}(\Phi(x)) = 1$. Then g is the inverse function of Φ.

We have

$$[x^n]f(g(x)) = \mathrm{Res}\,x^{-n-1}f(g(x))$$

$$= \mathrm{Res}\,\Phi(y)^{-n-1}\Phi'(y)f(y),$$

where we have made the substitution $x = \Phi(y)$ (so that $y = g(x)$) and used the Residue Composition Theorem.

For $n \neq 0$, we have

$$[x^n]f(g(x)) = -\frac{1}{n}[y^{-1}]f(y)\left(\Phi(y)^{-n}\right)'$$

$$= \frac{1}{n}[y^{-1}]f'(y)\Phi(y)^{-n}$$

$$= \frac{1}{n}[y^{n-1}]f'(y)\phi^n(y).$$

Here, in the second line, we have used the fact that

$$\mathrm{Res}(f'(x)g(x)) = -\mathrm{Res}(f(x)g'(x)),$$

a consequence of the fact that $\mathrm{Res}(f(x)g(x))' = 0$; in the third line we use the fact that $\Phi(x) = x/\phi(x)$.

For $n = 0$, we have

$$
\begin{aligned}
[x^0]f(g(x)) &= [y^0]f(y) - [y^{-1}]f(y)\phi'(y)\phi(y)^{-1} \\
&= f(0) + \mathrm{Res}(f'(y)\log(\phi(y)\phi^{-1}(0))),
\end{aligned}
$$

using the same principle as before and the fact that

$$(\log(\phi(y)\phi^{-1}(0)))' = \phi'(y)\phi(y)^{-1}.$$

Taking $f(x) = x$ in this result gives the form of Lagrange Inversion quoted earlier.

We proceed to an application, also taken from Goulden and Jackson, of the Residue Composition Theorem.

Example: a binomial identity We use the Residue Composition Theorem to prove that

$$\sum_{k=0}^{n} \binom{2n+1}{2k+1}\binom{j+k}{2n} = \binom{2j}{2n}.$$

We begin with the sum of the odd terms in $(1+x)^{2n+1}$:

$$\sum_{k=0}^{n} \binom{2n+1}{2k+1}x^{2k} = \frac{1}{2x}\left((1+x)^{2n+1} - (1-x)^{2n+1}\right).$$

Call the right-hand side of this equation $f(x)$. Now, if S is the sum that we want to evaluate, then

$$
\begin{aligned}
S &= [y^{2n}](1+y)^j \sum_{k=0}^{n} \binom{2n+1}{2k+1}(1+y)^k \\
&= \mathrm{Res}\, y^{-(2n+1)}(1+y)^j f((1+y)^{1/2}).
\end{aligned}
$$

Now we do the following rather strange thing: make the substitution $y = z^2(z^2 - 2)$. Then $\mathrm{val}(y(z)) = 2$, and $(1+y)^{1/2} = 1 - z^2$. So the Residue Composition Theorem gives

$$
\begin{aligned}
S &= \mathrm{Res}(z^2 - 1)^{2j}\left(\frac{1}{(z^2-2)^{2n+1}} - \frac{1}{z^{4n+2}}\right)z \\
&= \mathrm{Res}(z^2 - 1)^{2j}z^{-(4n+1)} \\
&= [z^{4n}](z^2 - 1)^{2j} \\
&= \binom{2j}{2n},
\end{aligned}
$$

as required. (In the second line we have used the fact that $(z^2 - 2)^{-(2n+1)}$ is a formal power series and so its residue is zero.)

12.2 Bernoulli numbers

We saw in Chapter 1 an asymptotic estimate for $n!$ which began by comparing $\log n! = \sum_{i=1}^{n} \log i$ to $\int_1^n \log x \, dx$. Obviously the comparison is not exact, but the approximation can often be improved by the Euler–Maclaurin sum formula. This formula involves the somewhat mysterious Bernoulli numbers, which crop up in a wide variety of other situations too.

The Bernoulli numbers B_n can be defined by the recurrence relation

$$B_0 = 1, \quad \sum_{k=0}^{n} \binom{n+1}{k} B_k = 0 \text{ for } n \geq 1.$$

Note that we can write the recurrence as

$$\sum_{k=0}^{n+1} \binom{n+1}{k} B_k = B_{n+1},$$

since the term B_{n+1} cancels from this equation (which expresses B_n in terms of earlier terms).

Conway and Guy, in *The Book of Numbers*, have a typically elegant presentation of the Bernoulli numbers. They write this relation as

$$(B+1)^{n+1} = B^{n+1}$$

for $n \geq 1$, where B^k is to be interpreted as B_k *after* the left-hand expression has been evaluated using the Binomial Theorem.

Thus,

$$B_2 + 2B_1 + 1 = B_2, \quad \text{whence} \quad B_1 = -\frac{1}{2},$$
$$B_3 + 3B_2 + 3B_1 + 1 = B_3, \quad \text{whence} \quad B_2 = \frac{1}{6},$$

and so on. Note that, unlike most of the sequences we have considered before, the Bernoulli numbers are not integers.

Theorem 12.5 *The exponential generating function for the Bernoulli numbers is*

$$\sum_{n \geq 0} \frac{B_n x^n}{n!} = \frac{x}{\exp(x) - 1}.$$

Proof Let $F(x)$ be the e.g.f., and consider $F(x)(\exp(x)-1)$. The coefficient of $x^{n+1}/(n+1)!$ is

$$(n+1)! \sum_{k=0}^{n} \left(\frac{B_k}{k!}\right) \left(\frac{1}{(n+1-k)!}\right) = \sum_{k=0}^{n} \binom{n+1}{k} B_k = 0$$

for $n \geq 1$. (Note that the sum runs from 0 to n rather than $n+1$ because we subtracted the constant term from the exponential.) The coefficient of x, however, is clearly 1. So the product is x.

Corollary 12.6 $B_n = 0$ *for all odd* $n > 1$.

Proof

$$F(x) + \frac{x}{2} = \frac{x}{2} \cdot \frac{\exp(x/2)+\exp(-x/2)}{\exp(x/2)-\exp(-x/2)} = \frac{x}{2} \coth\left(\frac{x}{2}\right)$$

which is an even function of x; so the coefficients of the odd powers of x are zero.

Corollary 12.7

$$B_n = \sum_{k=1}^{n} \frac{(-1)^k k! S(n,k)}{k+1}.$$

Proof Let $f(x) = \log(1+x)/x = \sum a_n x^n/n!$, where

$$a_n = \frac{(-1)^n n!}{(n+1)}.$$

By Theorem 3.12, $f(\exp(x)-1) = x/(\exp(x)-1) = \sum B_n x^n/n!$, where

$$B_n = \sum_{k=1}^{n} S(n,k) a_k.$$

One application of the Bernoulli numbers is in *Faulhaber's formula* for the sum of the kth powers of the first n natural numbers. Everyone knows that

$$\sum_{i=1}^{n} i = n(n+1)/2,$$

$$\sum_{i=1}^{n} i^2 = n(n+1)(2n+1)/6,$$

$$\sum_{i=1}^{n} i^3 = n^2(n+1)^2/4,$$

but how does the sequence continue?

Theorem 12.8

$$\sum_{i=1}^{n} i^k = \frac{1}{k+1} \sum_{j=0}^{k} \binom{k+1}{j} B_j (n+1)^{k+1-j}.$$

So, for example,

$$\begin{aligned}
\sum_{i=1}^{n} i^4 &= \frac{1}{5}\left((n+1)^5 - \frac{5}{2}(n+1)^4 + \frac{5}{3}(n+1)^3 - \frac{1}{6}(n+1) \right) \\
&= n(n+1)(6n^3 + 9n^2 + n - 1)/30.
\end{aligned}$$

Proof This argument is written out in the shorthand notation of Conway and Guy. Check that you can turn it into a more conventional proof!

We calculate

$$(n+1+B)^{k+1} - (n+B)^{k+1} = \sum_{j=1}^{k+1} \binom{k+1}{j} n^{k-j}((B+1)^j - B^j).$$

Now $(B+1)^j = B^j$ for all $j \geq 2$, so the only surviving term in this expression is

$$(k+1)n^k((B+1)^1 - B^1) = (k+1)n^k.$$

Thus we have

$$\frac{1}{k+1}((n+1+B)^{k+1} - (n+B)^{k+1}) = n^k,$$

from which by induction we obtain

$$\frac{1}{k+1}((n+1+B)^{k+1} - B^{k+1}) = \sum_{i=1}^{n} i^k.$$

The left-hand side of this expression is

$$\frac{1}{k+1} \sum_{j=0}^{k} \binom{k+1}{j} B_j (n+1)^{k+1-j},$$

as required.

Warning Conway and Guy use a non-standard definition of the Bernoulli numbers, as a result of which they have $B_1 = 1/2$ rather than $-1/2$. As a result, their formulae look a bit different.

How large are the Bernoulli numbers? The generating function $x/(\exp(x)-1)$ has a removable singularity at the origin; apart from this, the nearest singularities are at $\pm 2\pi i$, and so B_n is about $n!(2\pi)^{-n}$; in fact, it can be shown that

$$|B_n| = \frac{2n!\,\zeta(n)}{(2\pi)^n}$$

for n even, where $\zeta(n) = \sum_{k\geq 1} k^{-n}$. Of course, $B_n = 0$ if n is odd and $n > 1$.

Another curious formula for B_n is due to von Staudt and Clausen:

$$B_{2n} = N - \sum_{p-1 | 2n} \frac{1}{p}$$

for some integer N, where the sum is over the primes p for which $p-1$ divides $2n$.

12.2.1 Bernoulli polynomials

The *Bernoulli polynomials* $B_n(t)$ are defined by the formula

$$\frac{x \exp(tx)}{\exp(x) - 1} = \sum_{n\geq 0} \frac{B_n(t) x^n}{n!}.$$

Proposition 12.9 *The Bernoulli polynomials satisfy the following conditions:*

(a) $B_n(0) = B_n(1) = B_n$ *for* $n \neq 1$, *and* $B_1(0) = -1/2$, $B_1(1) = 1/2..$

(b) $B_n(t+1) - B_n(t) = nt^{n-1}$.

(c) $B'_n(t) = nB_{n-1}(t)$.

(d) $B_n(t) = \sum_{k=0}^{n} \binom{n}{k} B_{n-k} t^k$.

Proof All parts are easy exercises. Let $F(t) = x\exp(tx)/(\exp(x) - 1)$.
 (a) $F(0)$ is the e.g.f. for the regular Bernoulli numbers, and $F(1) = x + F(0)$.
 (b) $F(t+1) - F(t) = x\exp(tx)$.
 (c) $F'(t) = xF(t)$.
 (d) $F(t) = F(0)\exp(xt)$: use the rule for multiplying e.g.f.s.

The first few Bernoulli polynomials are

$$B_0(t) = 1, \qquad B_1(t) = t - \tfrac{1}{2}, \qquad B_2(t) = t^2 - t + \tfrac{1}{6},$$
$$B_3(t) = t^3 - \tfrac{3}{2}t^2 + \tfrac{1}{2}t, \qquad B_4(t) = t^4 - 2t^3 + t^2 - \tfrac{1}{30}.$$

12.3   The Euler–Maclaurin sum formula

Faulhaber's formula gives us an exact value for the sum of the values of a polynomial over the first n natural numbers. The Euler–Maclaurin formula generalises this to arbitrary well-behaved functions; instead of an exact value, we must be content with error estimates, which in some cases enable us to show that we have an asymptotic series.

The Euler–Maclaurin sum formula connects the sum

$$\sum_{i=1}^{n} f(i)$$

with the series

$$\int_{1}^{n} f(t)\,dt + \frac{1}{2}(f(1)+f(n)) + \sum \frac{B_{2i}}{(2i)!}\left(f^{(2i-1)}(n) - f^{(2i-1)}(1)\right),$$

where f is a 'sufficiently nice' function.

Here is a precise formulation due to de Bruijn.

Theorem 12.10 *Let f be a real function with continuous $(2k)$th derivative. Let*

$$S_k = \int_{1}^{n} f(t)\,dt + \frac{1}{2}(f(1)+f(n)) + \sum_{i=1}^{k} \frac{B_{2i}}{(2i)!}\left(f^{(2i-1)}(n) - f^{(2i-1)}(1)\right).$$

Then

$$\sum_{i=1}^{n} f(i) = S_k - R_k,$$

where the error term is

$$R_k = \int_{1}^{n} f^{(2k)}(t)\frac{B_{2k}(\{t\})}{(2k)!}\,dt,$$

with $B_{2k}(t)$ the Bernoulli polynomial and $\{t\} = t - \lfloor t \rfloor$ the fractional part of t.

Proof First let g be any function with continuous $(2k)$th derivative on $[0,1]$. We claim that

$$\frac{1}{2}(g(0)+g(1)) - \int_{0}^{1} g(t)\,dt$$

$$= \sum_{i=1}^{k} \frac{B_{2i}}{(2i)!}\left(g^{(2i-1)}(1) - g^{(2i-1)}(0)\right) - \int_{0}^{1} g^{(2k)}(t)\frac{B_{2k}(t)}{(2k)!}\,dt.$$

The proof is by induction: both the start of the induction (at $k = 1$) and the inductive step are done by integrating the last term by parts twice, using the fact that $B_n'(t) = nB_{n-1}(t)$ (see Proposition 12.9).

Now the result is obtained by applying this claim successively to the functions $g(x) = f(x+1)$, $g(x) = f(x+2)$, ..., $g(x) = f(x+n)$, and adding.

If f is a polynomial, then $f^{(2k)}(x) = 0$ for sufficiently large k, and the remainder term vanishes, giving Faulhaber's formula. For other applications, we must estimate the size of the remainder term.

There are various analytic conditions which guarantee a bound on the size of R_k, so that it can be shown that we have an asymptotic series for the sum. I will not give precise conditions here.

Example: Stirling's formula Let $f(x) = \log x$. Then $f^{(k)}(x) = \frac{(-1)^{k-1}(k-1)!}{x^k}$. We obtain the asymptotic series

$$c + n\log n - n + \frac{1}{2}\log n + \sum \frac{B_{2k}}{2k(2k-1)n^{2k-1}}$$

for

$$\sum_{i=1}^{n} \log i = \log n!.$$

The series begins $1/(12n) - 1/(360n^3) + 1/(1260n^5) + \cdots$. Exponentiating term-by-term (using the fact that, if $\log X = \log Y + o(n^{-k})$ then $X = Y(1 + o(n^{-k}))$), we obtain

$$n! \sim \sqrt{2\pi}\, \frac{n^{n+1/2}}{e^n} \left(1 + \frac{1}{12n} + \frac{1}{288n^2} + \cdots\right).$$

Note in passing that, for fixed n, this asymptotic series is divergent (see our earlier estimate for B_k).

Example: The harmonic series Applying Euler–Maclaurin to $f(x) = 1/x$, we get

$$\sum_{i=1}^{n} \frac{1}{i} \sim \log n + \gamma - \sum \frac{B_k}{kn^k},$$

where the sum begins $1/(2n) - 1/(12n^2) + 1/(120n^4) + \cdots$. Here γ is *Euler's constant*, a somewhat mysterious constant with value approximately $0.5772157\ldots$. Again the series is divergent for fixed n.

12.4 Poly-Bernoulli numbers

This is only a very brief introduction to these numbers, which were introduced by Masanobu Kaneko in 1997.

Kaneko gave the following definitions. Let

$$\mathrm{Li}_k(z) = \sum_{m=1}^{\infty} \frac{z^m}{m^k},$$

and let

$$\frac{\mathrm{Li}_k(1-\mathrm{e}^{-x})}{1-\mathrm{e}^{-x}} = \sum_{n=0}^{\infty} B_n^{(k)} \frac{x^n}{n!}.$$

The numbers $B_n^{(k)}$ are the *poly-Bernoulli numbers of order k*.

Kaneko gave a couple of nice formulae for the poly-Bernoulli numbers of negative order, of which one is relevant here:

Theorem 12.11 (Kaneko)

$$B_n^{(-k)} = \sum_{j=0}^{\min(n,k)} (j!)^2 S(n+1,j+1) S(k+1,j+1).$$

This formula has the (entirely non-obvious) corollary that these numbers have a symmetry property: $B_n^{(-k)} = B_k^{(-n)}$ for all non-negative integers n and k.

Now we have a somewhat unexpected connection with a counting problem: the poly-Bernoulli numbers of negative order evaluate the chromatic polynomial of complete bipartite graphs at -1 (see Theorem 9.3). This application is taken from an unpublished paper of the author with Celia Glass and Robert Schumacher.

Theorem 12.12 *The number of acyclic orientations of K_{n_1,n_2} is $B_{n_1}^{(-n_2)}$.*

Proof We verify Kaneko's formula above.

Let A and B be the two bipartite blocks; we will imagine their vertices as coloured amber and blue respectively. Now any acyclic orientation of the graph can be obtained by ordering the vertices and making the edges point from smaller to greater. If we do this, we will have alternating amber and blue intervals; the ordering within each interval is irrelevant, but the ordering of the intervals themselves matters.

In terms of structure, call two points $a_1, a_2 \in A$ *equivalent* if the orientations of $\{a_1, b\}$ and $\{a_2, b\}$ are the same for all $b \in B$. Points are equivalent if and only if they are not separated by a point of B in any ordering giving rise to the acyclic orientation. Similarly for B. This gives us the intervals, and clearly they are interleaved.

To get around the problem that the first point in the ordering might be in either A or B, and similarly for the last point, we use the following trick. Add a dummy amber point a_0 to A and a dummy blue point b_0 to B. Now partition $A \cup \{a_0\}$ and $B \cup \{b_0\}$ into the same number, say k, of parts. This can be done in $S(n_1 + 1, k) S(n_2 + 1, k)$ ways. Now we order the parts so that

- the part containing a_0 is first;
- the colours alternate;
- the part containing b_0 is last.

This can be done in $(k-1)!^2$ ways. Finally, delete the dummy points.

Summing over k gives the total number claimed.

12.5   Hayman's Theorem

A number of non-trivial analytic results have been proved for the purpose of obtaining asymptotic formulae for combinatorially defined numbers. These include theorems of Hayman, Meir and Moon, and Bender. I will not give proofs of these theorems, but treat them as black boxes and give examples to illustrate their use.

Hayman's Theorem is an important result on the asymptotic behaviour of the coefficients of certain *entire* functions (i.e., functions which are analytic in the entire complex plane).

The theorem applies only to a special class of such functions, the so-called *H-admissible* or *Hayman-admissible* functions. Rather than attempt to give a general definition of this class, I will state a theorem of Hayman showing that it is closed under certain operations, which suffice to show that any function in which we are interested is H-admissible. See Hayman's paper in the bibliography, or Odlyzko's survey.

Theorem 12.13   *(a) If f is H-admissible and p is a polynomial with real coefficients, then $f + p$ is H-admissible.*

(b) If p is a non-constant polynomial with real coefficients such that $\exp(p(x)) = \sum q_n x^n$ with $q_n > 0$ for $n \geq n_0$, then $\exp(p(x))$ is H-admissible.

(c) If p is a non-constant real polynomial with leading term positive, and f is H-admissible, then $p(f(x))$ is H-admissible.

(d) If f and g are H-admissible, then $\exp(f(x))$ and $f(x)g(x)$ are H-admissible.

Corollary 12.14   *The exponential function is H-admissible.*

Now *Hayman's Theorem* is the following.

Theorem 12.15   *Let $f(x) = \sum_{n\geq0} f_n x^n$ be H-admissible. Let $a(x) = xf'(x)/f(x)$ and $b(x) = xa'(x)$, and let r_n be the smallest positive root of the equation $a(x) = n$. Then*

$$f_n \sim \frac{1}{\sqrt{2\pi b_n}} f(r_n) r_n^{-n}.$$

Example: Stirling's formula   Take $f(x) = \exp(x)$ (we have noted that this function is admissible), so that $f_n = 1/n!$. Now $a(x) = x = b(x)$, and $r_n = n$. Thus

$$\frac{1}{n!} = \frac{1}{\sqrt{2\pi n}} e^n n^{-n},$$

which is just Stirling's formula the other way up!

Example: Bell numbers   Let $f(x) = \exp(\exp(x) - 1)$, so that $f_n = B(n)/n!$, where $B(n)$ is the number of partitions of an n-set. This function is H-admissible. Now $a(x) = xe^x$ and $b(x) = (x + x^2)e^x$.

The number r_n is the smallest positive solution of $xe^x = n$. In terms of this, we have

$$\frac{B(n)}{n!} \sim \frac{1}{\sqrt{2\pi n(1 + r_n)}} e^{n/r_n - 1} r_n^{-n},$$

and so by Stirling's formula,

$$B(n) \sim \frac{1}{\sqrt{1 + r_n}} \left(\frac{n}{er_n}\right)^n e^{n/r_n - 1}.$$

Of course, this is not much use without a good estimate for r_n. However, for $n = 100$, the right-hand side is within 0.4% of $B(100)$.

In fact, it can be shown that

$$r_n = \log n - \log\log n + O\left(\frac{\log\log n}{\log n}\right),$$

from which it can be deduced that

$$\log B(n) \sim n\log n - n\log\log n - n.$$

12.6   Theorems of Meir and Moon and of Bender

The theorem of Meir and Moon (which has been generalised by Bender) gives the asymptotics of the coefficients of a power series defined by Lagrange inversion. Typically we have to find the inverse function of f. Setting $\phi(x) = x/f(x)$, the inverse function g is given by the functional equation $g(y) = y\phi(g(y))$. Replacing y by x and g by f, the theorem is as follows.

Theorem 12.16 *Let $y = f(x) = \sum f_n x^n$ satisfy the equation*

$$y = x\Phi(y),$$

where Φ is analytic in some neighbourhood of the origin, with $\Phi(x) = \sum a_n x^n$. Suppose that the following conditions hold:

(a) $a_0 = 1$ and $a_n \geq 0$ for $n \geq 0$.

(b) $\gcd\{n : a_n > 0\} = 1$.

(c) There is a positive real number α, inside the circle of convergence of Φ, satisfying

$$\alpha\Phi'(\alpha) = \Phi(\alpha).$$

Then

$$f_n \sim Cn^{-3/2}\beta^n,$$

where $C = \sqrt{\Phi(\alpha)/2\pi\Phi''(\alpha)}$ and $\beta = \Phi(\alpha)/\alpha = \Phi'(\alpha)$.

Example: Rooted trees The generating function $y = T^*(x)$ for labelled rooted trees satisfies

$$y = x\exp(y).$$

The exponential function converges everywhere, and the solution of $\alpha\exp(\alpha) = \exp(\alpha)$ is clearly $\alpha = 1$, so that $\beta = e$ and $C = \sqrt{1/2\pi}$. Hence the number T_n^* of labelled rooted trees on n vertices satisfies

$$\frac{T_n^*}{n!} = \frac{1}{\sqrt{2\pi}} n^{-3/2} e^n.$$

Since $T_n^* = n^{n-1}$ by Cayley's Theorem, we obtain

$$n! \sim \sqrt{2\pi} \frac{n^{n+1/2}}{e^n},$$

in other words, Stirling's formula.

Bender's Theorem generalises the theorem of Meir and Moon by treating a very much more general class of implicitly defined functions. Thus, y will be defined as a function of x by the equation $F(x,y) = 0$. In the case of Meir and Moon, we have $F(x,y) = y - x\Phi(y)$.

Theorem 12.17 *Suppose that the function $y = f(x)$ is defined implicitly by the equation $F(x,y) = 0$, and let $f(x) = \sum_{n\geq 0} f_n x^n$. Suppose that there exist real numbers ξ and η such that*

(a) F is analytic in a neighbourhood of (ξ, η);

(b) $F(\xi, \eta) = 0$ and $F_y(\xi, \eta) = 0$, but $F_x(\xi, \eta) \neq 0$ and $F_{yy}(\xi, \eta) \neq 0$ (subscripts denote partial derivatives);

(c) the only solution of $F(x,y) = F_y(x,y) = 0$ with $|x| \leq \xi$ and $|y| \leq \eta$ is $(x,y) = (\xi, \eta)$.

Then

$$f_n \sim Cn^{-3/2}\xi^{-n},$$

where

$$C = \sqrt{\frac{\xi F_x(\xi, \eta)}{2\pi F_{yy}(\xi, \eta)}}.$$

Example: Wedderburn–Etherington numbers   Recall from Chapter 4 that the generating function for these numbers satisfies

$$f(x) = x + \frac{1}{2}(f(x)^2 + f(x^2)).$$

Here we have $F(x,y) = y - x - (y^2 + g(x))/2$, where $g(x) = f(x^2)$, which we regard as a 'known' function (using a truncation of its Taylor series to approximate it).

The equation $F_y(\xi, \eta) = 0$ gives us that $\eta = 1$; the equation $F(\xi, \eta) = 0$ then gives $g(\xi) = 1 - 2\xi$. This equation can be solved numerically (it is the same one we solved in Chapter 4 to find the radius of convergence of $f(x)$). The remaining conditions of the theorem can then be verified.

We obtain $\xi^{-1} = 2.483\ldots$, and hence

$$f_n \sim Cn^{-3/2}\xi^{-n},$$

where C can also be found numerically if desired.

Remark   This asymptotic formula illustrates a common occurrence both in enumerative combinatorics and in related areas such as statistical mechanics. We have an asymptotic formula of the shape $cn^{-r}\alpha^n$, where the exponent r is known precisely, but the exponential constant α (which has a much greater effect on the asymptotics) is known only approximately.

12.7   Exercises

12.1   (a) Let f be a formal power series with constant term zero and coefficient of x equal to 1. Suppose that all coefficients of f are integers. Show that the unique formal power series g such that $g(f(x)) = x$ has all its coefficients integers.

(b) Show that the preceding statement holds if we replace the integers by an arbitrary commutative ring with identity.

(c) Let R be a commutative ring with identity, and let $\mathscr{N}(R)$ denote the set of formal power series $\sum a_n x^n$ in $R[[x]]$ with $a_0 = 0$ and $a_1 = 1$. Show that $\mathscr{N}(R)$, with the operation of composition, is a group.

(d) Let $R = \mathbb{Z}$. Find the inverse of $x - x^2$ in $\mathscr{N}(R)$.

Remark   The group $\mathscr{N}(R)$ is known to group theorists as the *Nottingham group* over R.

12.2 Let s_n be the number of involutions in S_n (permutations of $\{1,\ldots,n\}$ whose square is the identity). We showed in Chapter 4 that

$$\sum_{n \geq 0} \frac{s_n x^n}{n!} = \exp\left(x + \frac{x^2}{2}\right).$$

Use Hayman's Theorem to show that

$$s_n \sim \frac{1}{\sqrt{2}} \left(\frac{n}{e} \right)^{n/2} e^{\sqrt{n}-1/4}.$$

12.3 Let $W(x)$ be the generating function for the Catalan numbers, shifted back one place:

$$W(x) = \sum_{n \geq 0} C_{n+1} x^n.$$

Put $W(x) = 1 + u(x)$. Show that

$$u(x) = x(1 + u(x))^2.$$

Apply Lagrange inversion to show that

$$C_{n+1} = \frac{1}{n} \binom{2n}{n-1}$$

for $n \geq 1$. Reconcile this with the formula for Catalan numbers found in Chapter 4.

CHAPTER 13

Bibliography and further directions

In this chapter I list a few books, papers and websites which may be useful if you would like to follow up some of the things I have discussed.

13.1 The On-line Encyclopedia of Integer Sequences

The On-line Encyclopedia of Integer Sequences, available at the URL

https://oeis.org/,

is an essential resource for anyone doing research in combinatorics. For example, suppose you are trying to count the number of arrangements of n zeros and ones around a circle in which no two ones are consecutive (Exercise 2.8(b)). You might reasonably assume that $n \geq 3$, and calculate that for $n = 3, 4, 5$ there are respectively 4, 7 and 11 such arrangements. If you type these three numbers into the Encyclopedia, you find many matches (I found 521 when I tried it on 24 October 2016), but near the top (indeed, at the top when I did the experiment) is an entry for the Lucas numbers. The entry gives a recurrence relation (identical to that for Fibonacci numbers), congruences modulo primes, representation in terms of hyperbolic functions, and much more, including (most importantly) ten references to the literature, short programs for computing the numbers in various programming languages, further web links, open problems, and cross-references to related sequences. Now you can either prove directly that the numbers you are interested in satisfy the Fibonacci recurrence (and hence coincide with the Lucas numbers), or check in the literature for further information which will help you make the identification.

On the Encyclopedia's website, you will find different ways of viewing the sequence and information about it, pointers to interesting or mysterious sequences, a formula for the terms of the sequence or its generating function if known, and several articles by the editor Neil Sloane and others describing uses of the Encyclopedia in research. I have used it myself on a number of occasions.

13.2 Books on combinatorial enumeration

Here are the books referred to in the text.

F. Bergeron, G. Labelle and P. Leroux, *Combinatorial Species and Tree-like Structures*, Encyclopedia of Mathematics and its Applications **67**, Cambridge University Press, Cambridge, 1998.

This is the standard reference on species.

P. J. Cameron, *Oligomorphic Permutation Groups*, London Math. Soc. Lecture Notes **152**, Cambridge University Press, Cambridge, 1990.

The book that introduced oligomorphic permutation groups, and describes their links with combinatorics and logic as well as group theory.

P. J. Cameron, *Permutation Groups*, London Math. Soc. Student Texts **45**, Cambridge University Press, Cambridge, 1999.

Background material for Chapters 7 and 10.

J. H. Conway and R. K. Guy, *The Book of Numbers*, Springer-Verlag, New York, 1996.

I followed the treatment of Bernoulli numbers in this book, but there is much else here to delight the combinatorial enumerator.

P. Flajolet and R. Sedgewick, *Analytic Combinatorics*, Cambridge University Press, Cambridge, 2009.

This is the most comprehensive modern account of recent developments in analytic methods for studying combinatorial sequences via generating functions.

R. Fraïssé, *Theory of Relations*, North-Holland, Amsterdam, 1986.

A wide-ranging account of relational structures, including material on homogeneous structures relevant to Chapter 10.

M. R. Garey and D. S. Johnson, *Computers and Intractability: An Introduction to the Theory of* NP-*completeness*, W. H. Freeman, San Francisco, 1979.

The standard reference on computational complexity.

I. P. Goulden and D. M. Jackson, *Combinatorial Enumeration*, Wiley-Interscience, New York, 1983.

A very complete account of much of the material covered in this book.

Wilfrid Hodges, *A Shorter Model Theory*, Cambridge University Press, Cambridge, 1997.

Includes a good discussion of countably (and uncountably) categorical theories, alluded to under oligomorphic permutation groups in Chapter 10.

V. Kac and P. Cheung, *Quantum Calculus*, Springer, New York, 2002.

Learn more about the q-calculus and h-calculus in this very accessible account.

Sergey Kitaev, *Patterns in permutations and words*, Monographs in Theoretical Computer Science, Springer-Verlag, 2011.

Permutation patterns were discussed in connection with Catalan numbers in Chapter 4. This is a recent survey.

J. H. van Lint and R. M. Wilson, *A Course in Combinatorics*, Cambridge University Press, Cambridge, 2001.

A good introduction to many topics in combinatorics. In particular, it contains an accessible account of Egoritsjev's proof of the van der Waerden conjecture.

I. G. Macdonald, *Symmetric functions and Hall polynomials*, Oxford University Press, Oxford, 1999.

The theory of symmetric functions, diagrams and Young tableaux is at the centre of a beautiful area of mathematics ranging from orthogonal polynomials to representations of algebraic groups. This book was awarded the American Mathematical Society's Steele Prize for Mathematical Exposition in 2009.

Marko Petkovsek, Herbert S. Wilf and Doron Zeilberger, $A = B$, A. K. Peters Ltd., Wellesley, MA, 1996.

This book describes a remarkable advance in the automated proving of binomial coefficient identities and similar results. You should look at the web page http://www.math.upenn.edu/~wilf/AeqB.html where the book is free to download.

A. Slomson, *An Introduction to Combinatorics*, Chapman and Hall, London, 1991.

An introductory textbook, not restricted to enumerative combinatorics; I have based my account of Stirling's formula on Slomson's.

R. P. Stanley, *Enumerative Combinatorics* (2 volumes), Cambridge University Press, Cambridge, 2000, 2001.

Particularly strong and detailed on the algebraic aspects of enumerative combinatorics, including the Robinson–Shensted–Knuth algorithm and its implications.

Another important source is the two-volume *Handbook of Combinatorics* (ed. R. L. Graham, M. Grötschel and L. Lovasz), North-Holland, Amsterdam, 1995. This book covers all aspects of combinatorics, not just enumeration; but the 167-page article on Asymptotic Enumeration Methods by Andrew Odlyzko is essentially a monograph in its own right. There is also a very relevant article on Algebraic Enumeration by Ira M. Gessel and Richard P. Stanley.

13.3   Papers cited in the text

E. A. Bender, Asymptotic methods in enumeration, *SIAM Review* **16** (1974), 485–515.

A. T. Benjamin, G. M. Levin, K. Mahlburg and J. J. Quinn, Random approaches to Fibonacci identities, *Amer. Math. Monthly* **107** (2000), 511–516.

N. Boston, W. Dabrowski, T. Foguel, P. J. Gies, J. Leavitt, D. T. Ose and D. A. Jackson, The proportion of fixed-point-free elements of a transitive permutation group, *Commun. Algebra* **21** (1993), 3259–3275.

P. J. Cameron, Oligomorphic permutation groups, in *Perspectives in Mathematical Sciences II* (ed. N. S. Narasimha Sastry, T. S. S. R. K. Rao, Mohan Delampady and B. Rajeev), World Scientific, Singapore, 2009, pp. 37–61.

P. J. Cameron and A. M. Cohen, On the number of fixed point free elements of a permutation group, *Discrete Math.* **106/107** (1992), 135–138.

P. J. Cameron, D. Gewurz and F. Merola, Product action, *Discrete Math.* **308** (2008), 386–394.

P. J. Cameron, B. Jackson and J. D. Rudd, An orbital Tutte polynomial for graphs and matroids, *Discrete Math.* **308** (2008), 920–930.

P. J. Cameron, T. Prellberg and D. Stark, Asymptotic enumeration of incidence matrices, *Journal of Physics: Conference Series* **42** (2006), 59–70.

A. Dress and T. Müller, Decomposable functors and the exponential principle, *Advances in Mathematics* **129** (1997), 188–221.

S. Eberhard, K. Ford and B. Green, Permutations fixing a k-set, `https://arxiv.org/abs/1507.04465`

W. T. Gowers, The two cultures of mathematics, pp. 65–78 in *Mathematics: Frontiers and Perspectives* (ed. V. Arnold, M. Atiyah, P. Lax and B. Mazur), American Math. Soc., Providence, 1999.

L. J. Guibas and A. M. Odlyzko, String overlaps, pattern matching, and nontransitive games, *J. Combinatorial Theory* (A) **30** (1981), 183–208.

W. K. Hayman, A generalization of Stirling's formula, *J. Reine Angew. Math.* **196** (1956), 67–95.

A. Joyal, Une theorie combinatoire des séries formelles, *Advances in Math.* **42** (1981), 1–82.

M. Kaneko, Poly-Bernoulli numbers, *Journal de Théorie des Nombres de Bordeaux* **9** (1997), 221–228.

A. Laradji and A. Umar, Combinatorial results for the symmetric inverse semigroup, *Semigroup Forum* **75** (2007), 221–236.

R. Lewis and S. P. Norton, On a problem raised by P. J. Cameron, *Discrete Mathematics* **138** (1995), 315–318.

N. Linial and N. Nisan, Approximate inclusion-exclusion, *Combinatorica* **10** (1990), 349–365.

S. Majid, Free braided differential calculus, braided binomial theorem and the braided exponential map, *J. Math. Phys.* **34** (1993), 4843–4856.

F. Merola, Orbits on n-tuples for infinite permutation groups, *Europ. J. Combinatorics* **22** (2001), 225–241.

Peter M. Neumann, A lemma that is not Burnside's, *The Mathematical Scientist* **4** (1979), 133–141.

Maurice Pouzet, When is the orbit algebra of a group an integral domain? Proof of a conjecture of P. J. Cameron, *Theor. Inform. Appl.* **42** (2008), 83–103.

A. D. Sokal, The multivariate Tutte polynomial (alias Potts model) for graphs and matroids, in *Surveys in Combinatorics 2005* (ed. B. S. Webb), London Math. Soc. Lecture Notes, **327**, Cambridge Univ. Press, Cambridge, 2005, pp. 173–226.

R. P. Stanley, Acyclic orientations of graphs, *Discrete Math.* **5** (1973), 171–178.

H. S. Wilf, The 'Snake Oil' method for proving combinatorial identities, in *Surveys in Combinatorics* (ed. J. Siemons), London Math. Soc. Lecture Note Series **141**, Cambridge University Press, Cambridge, 1989, pp. 208–217.

Doron Zeilberger, Automatic counting of tilings of skinny plane regions, in *Surveys in Combinatorics 2013* (ed. Simon R. Blackburn, Stefanie Gerke and Mark Wildon), London Math. Soc. Lecture Note Series **409**, Cambridge University Press, Cambridge, 2013, pp. 363–378.

Index

abelian group, 142, 160
active form, 42
acyclic orientations, 158, 206
addition
    of formal power series, 17
age, 179
alternating group, 43, 59, 104
amalgamation property, 179
Appel, K., 158
approximate Inclusion–Exclusion, 155
associate, 163
asymptotic estimate, 2, 187
    for Bell numbers, 208
    for factorials, 5, 188, 205, 207
    for involutions, 210
    for partition numbers, 6
    for Wedderburn–Etherington
        numbers, 84, 210
asymptotic series, 188
    for factorials, 205
Atiyah, M., x
Axiom of Choice, 145

ballot numbers, 93
Bell numbers, 6, 38, 80, 208
Bender's Theorem, 209
Bender, E. A., 209, 215
Benjamin, A. T., 215
Bergeron, F., 178, 213
Bernoulli numbers, 125, 200
Bernoulli polynomials, 203

Bezembinder, T., 60
bijective proof, x, 57, 59, 61, 89
binary tree, 89
binomial coefficients, 4, 24, 29
    signed, 31
binomial series, 24
Binomial Theorem, 4, 24, 30, 33, 59, 68
Birkhoff's Theorem, 109
Birkhoff, G. D., 156
blocks, 183
Boston, N., 215
braided categories, 115

C-finite, 73
Cantor's theorem, 181
Catalan numbers, 83, 210, 211
categories, 178
    braided, 115
Cauchy–Binet formula, 167
Cauchy–Schwarz inequality, 10
Cayley's Theorem, 168, 170, 176
Central Limit Theorem, 37, 58
characteristic polynomial, 66
Cheung, P., 214
chromatic polynomial, 157
clinical trials, 61
Cohen, A. M., 215
colouring, proper, 157
commutation relation, 119
commutative ring with identity, 17
complete symmetric functions, 126